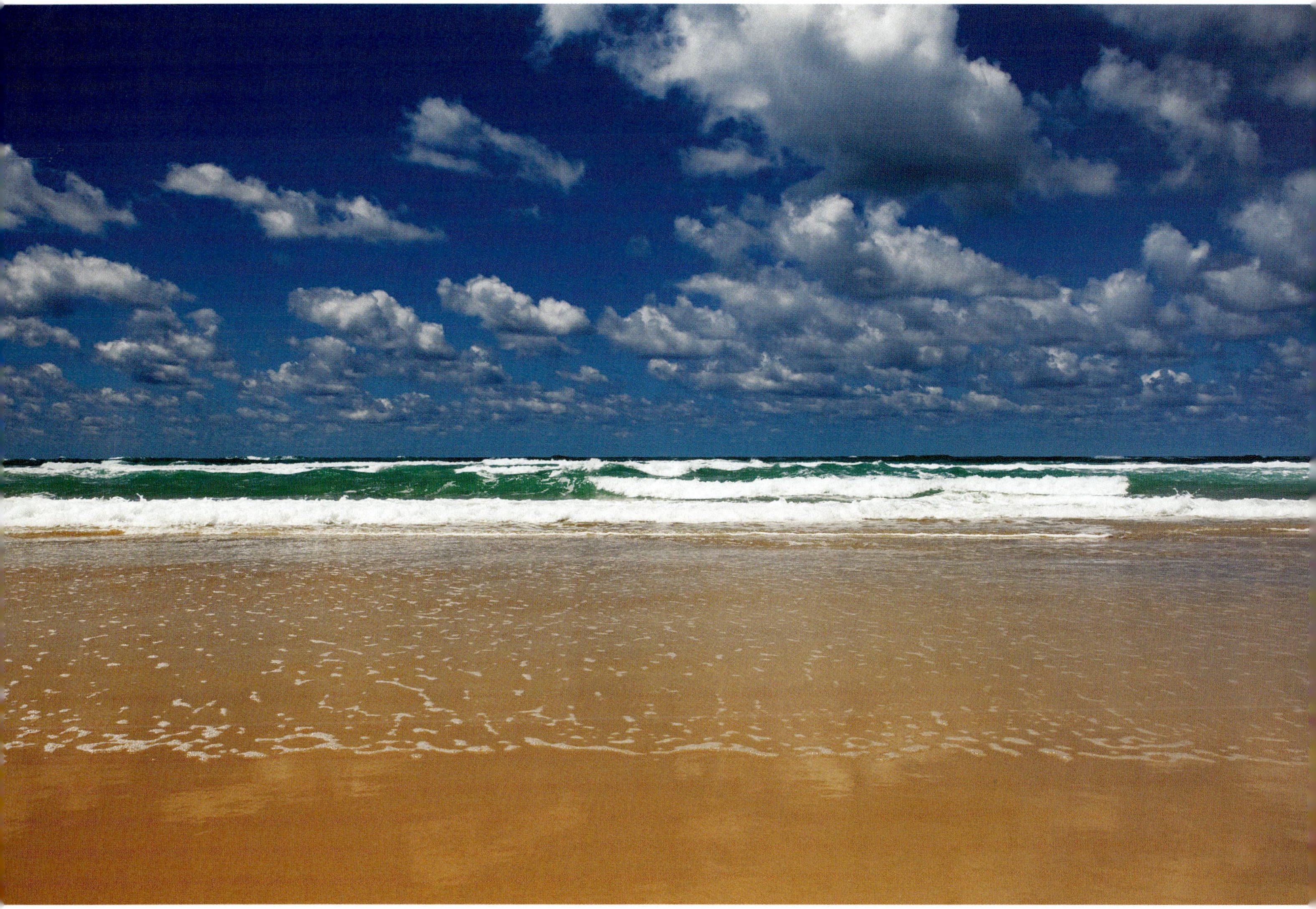

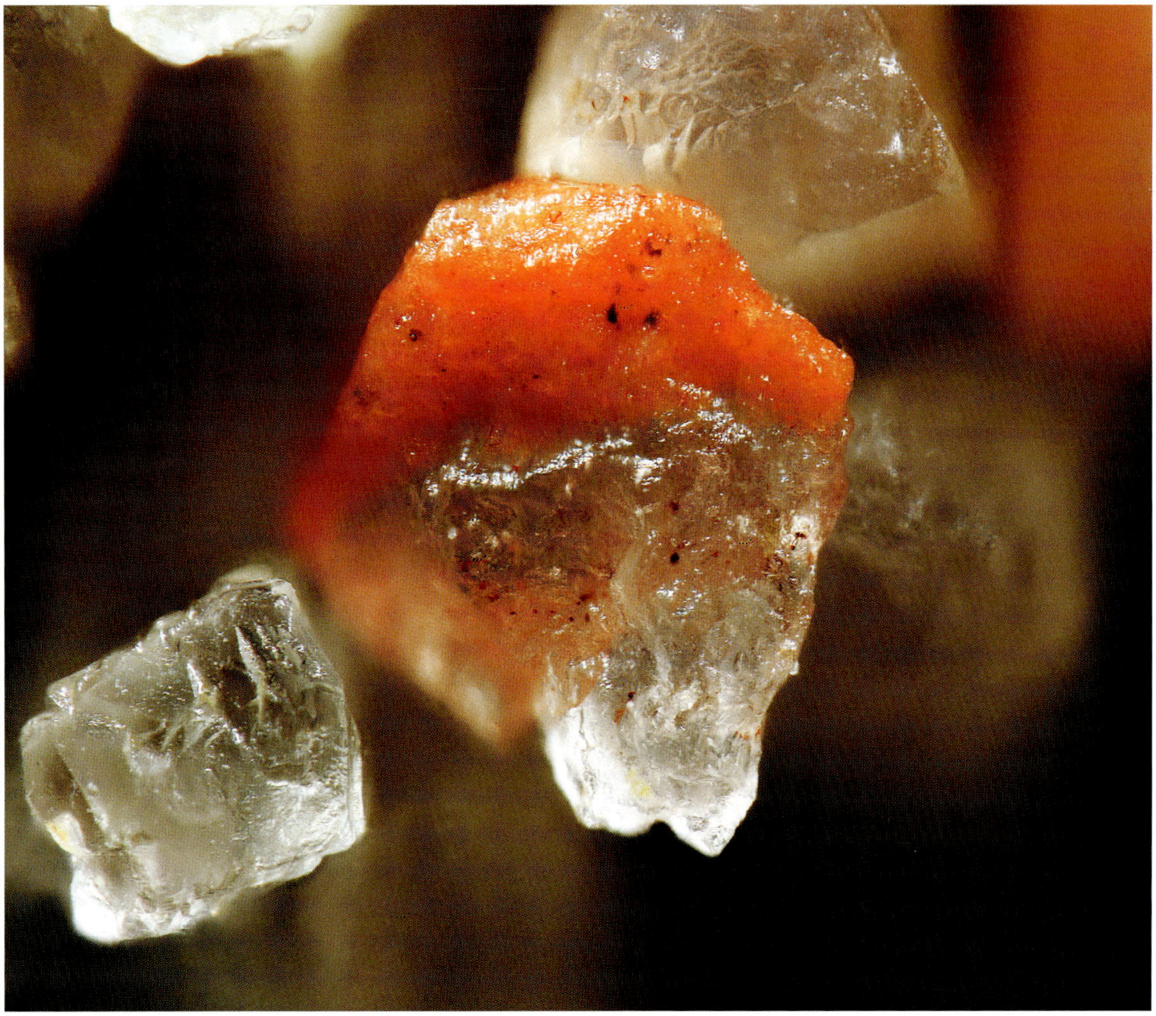

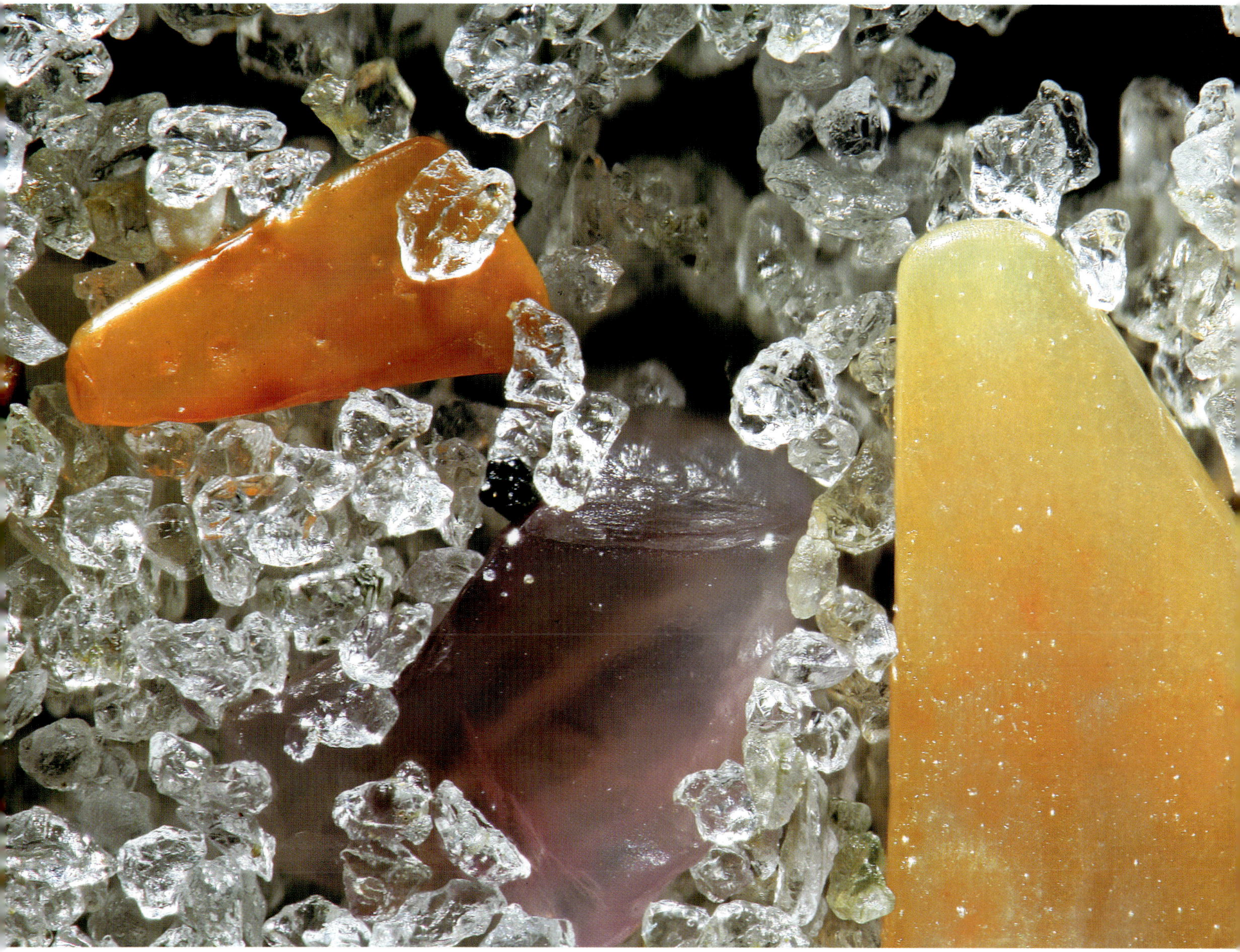

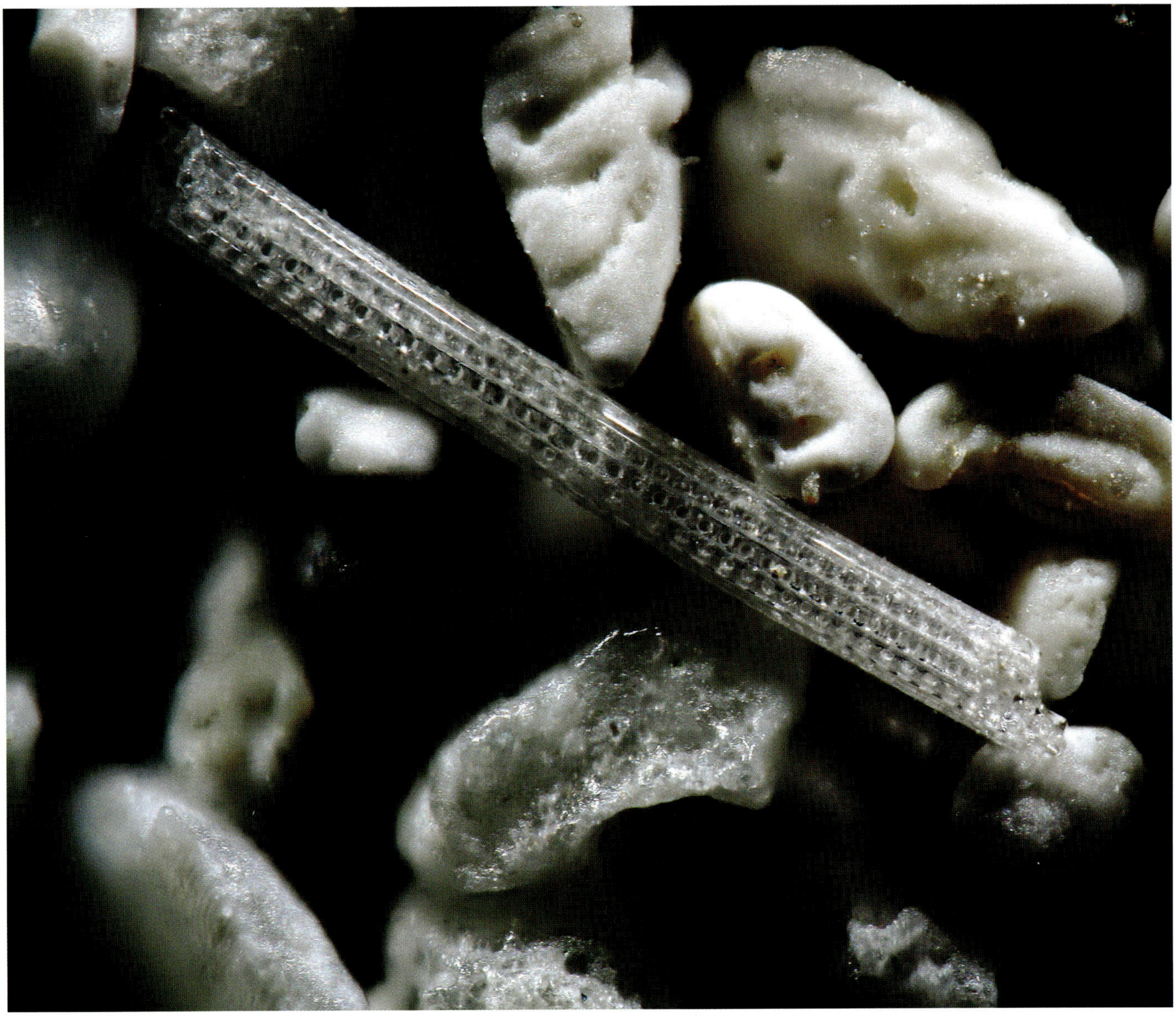

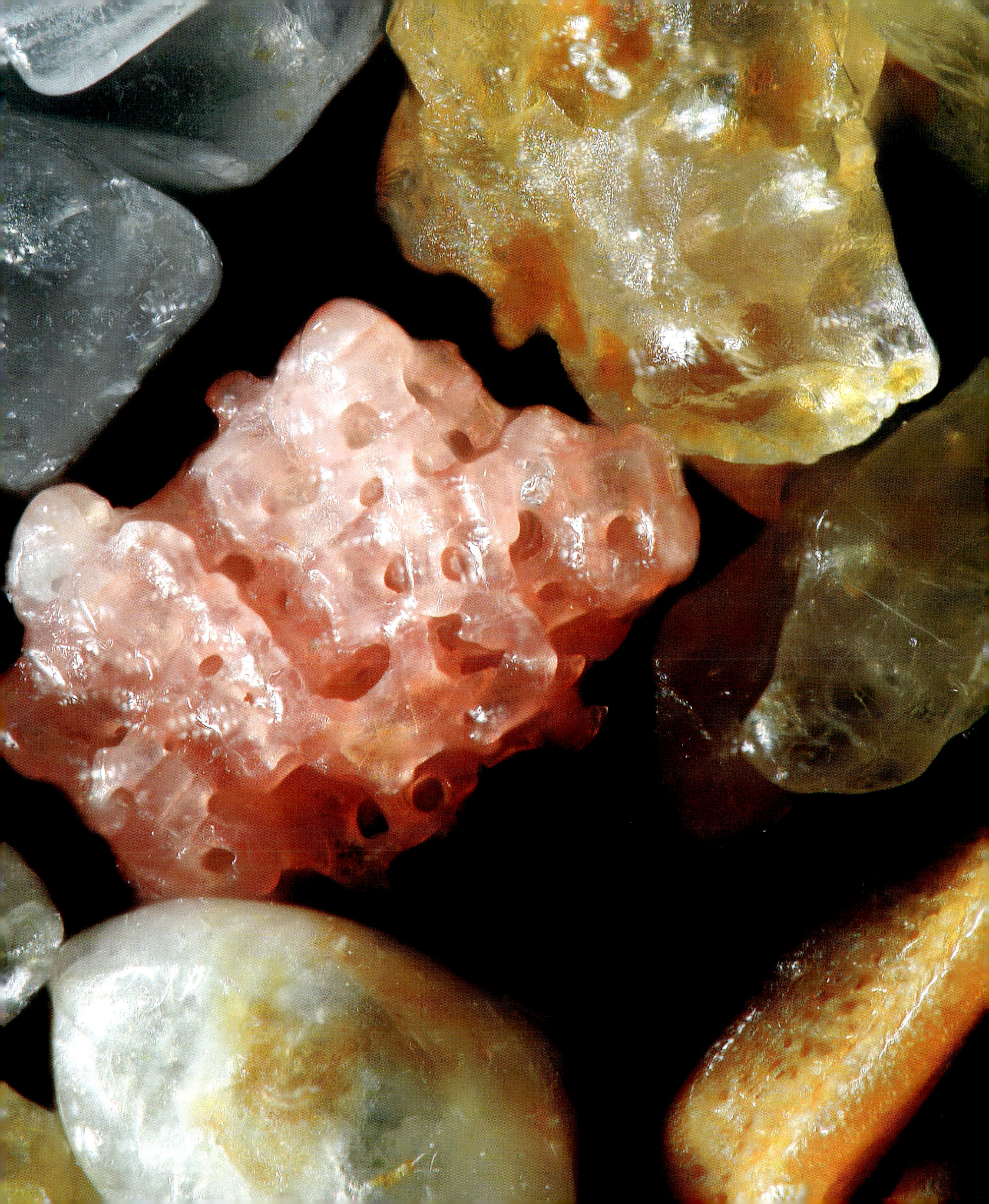

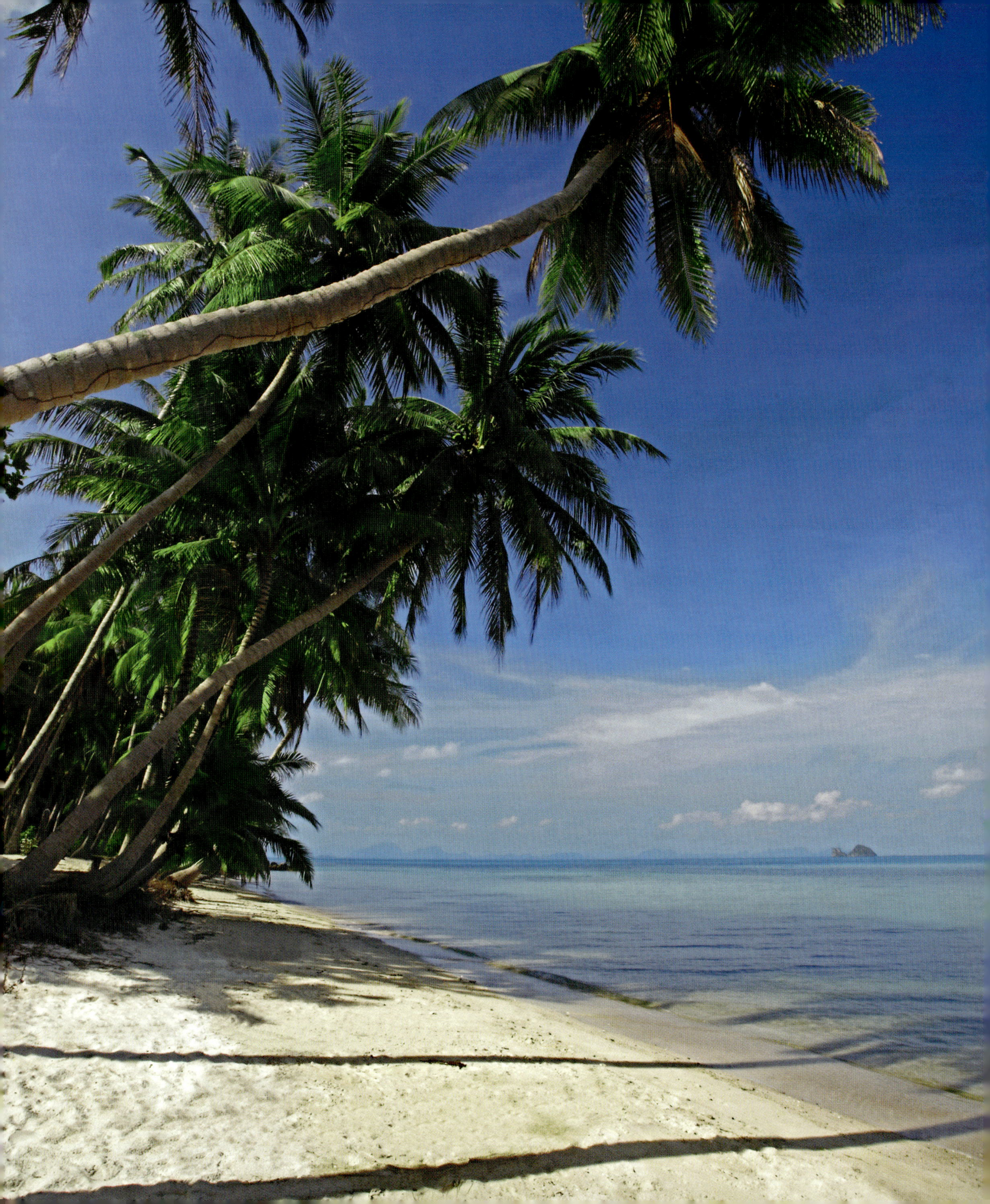

a *Grain* of SAND

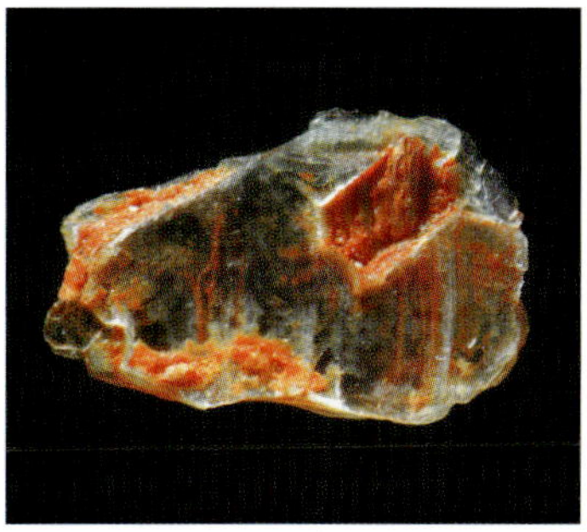
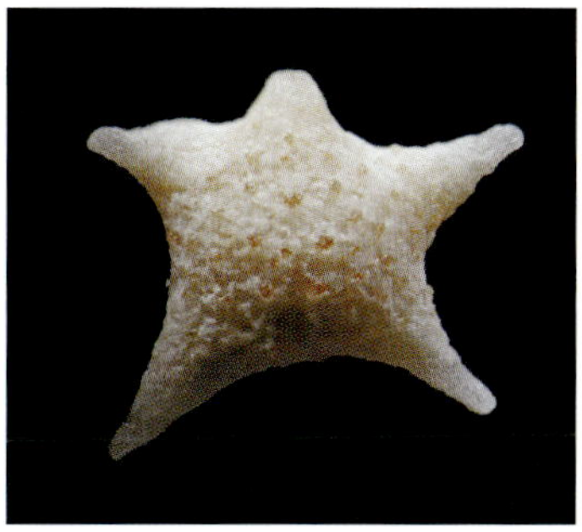
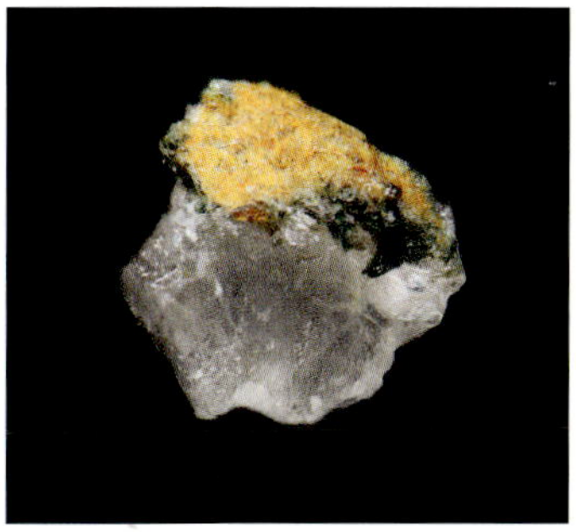

Text and photography by
Dr. Gary Greenberg

Voyageur Press

Voyageur Press, an imprint of MBI Publishing Company LLC, 400 First Ave. North Suite 300, Minneapolis, MN 55401 USA

Voyageur Press titles are also available at discounts in bulk quantity for industrial or sales-promotional use. For details write to Special Sales Manager at MBI Publishing Company, 400 First Ave. North Suite 300, Minneapolis, MN 55401 USA.

To find out more about our books, join us online at www.voyageurpress.com.

Library of Congress Cataloging-in-Publication Data

Greenberg, Gary, 1944-
A grain of sand : nature's secret wonder / text and photography by Gary Greenberg.
p. cm.
Includes index.
ISBN-13: 978-0-7603-3198-9 (plc w/ jacket)
1. Microphotography. 2. Sand. I. Title.
TR835.G74 2008
779'.36—dc22

2007034731

Edited by Danielle J. Ibister
Designed by Maria Friedrich

Printed in China

On the cover, main: Maui beach. *Stephanie Coffman, Shutterstock*

On the cover, insets (left to right): Purple sea urchin spine tip; blue shell fragment; olivine; heart-shaped shell fragment.

On the back cover (left to right): Mineral grain; pink coral; star-shaped foram.

Page 1: Australian beach. Peter *G, Shutterstock*

Page 2: A sand grain from Eilat, Israel (magnification 150x).

Page 3: Brightly colored sand grains among quartz from Flagler Beach, Florida (magnification 200x).

Page 4: A delicate sea urchin spine from Utila, Honduras (magnification 165x).

Page 5: Pink coral and shell fragments from Geriba Beach, Brazil (magnification 200x).

Page 6: White sand on a tropical beach. *Ersler Dmitry, Shutterstock*

Facing page: A mixture of biological and mineral sand from Makena Big Beach, Maui, Hawaii (magnification 300x).

Page 10: Sand grains from the Thames River, London (magnification 300x).

Page 13: The tip of a spiral shell (magnification 250x).

Page 16: Colorful mineral sand grains from Umhlanga, South Africa (magnification 250x).

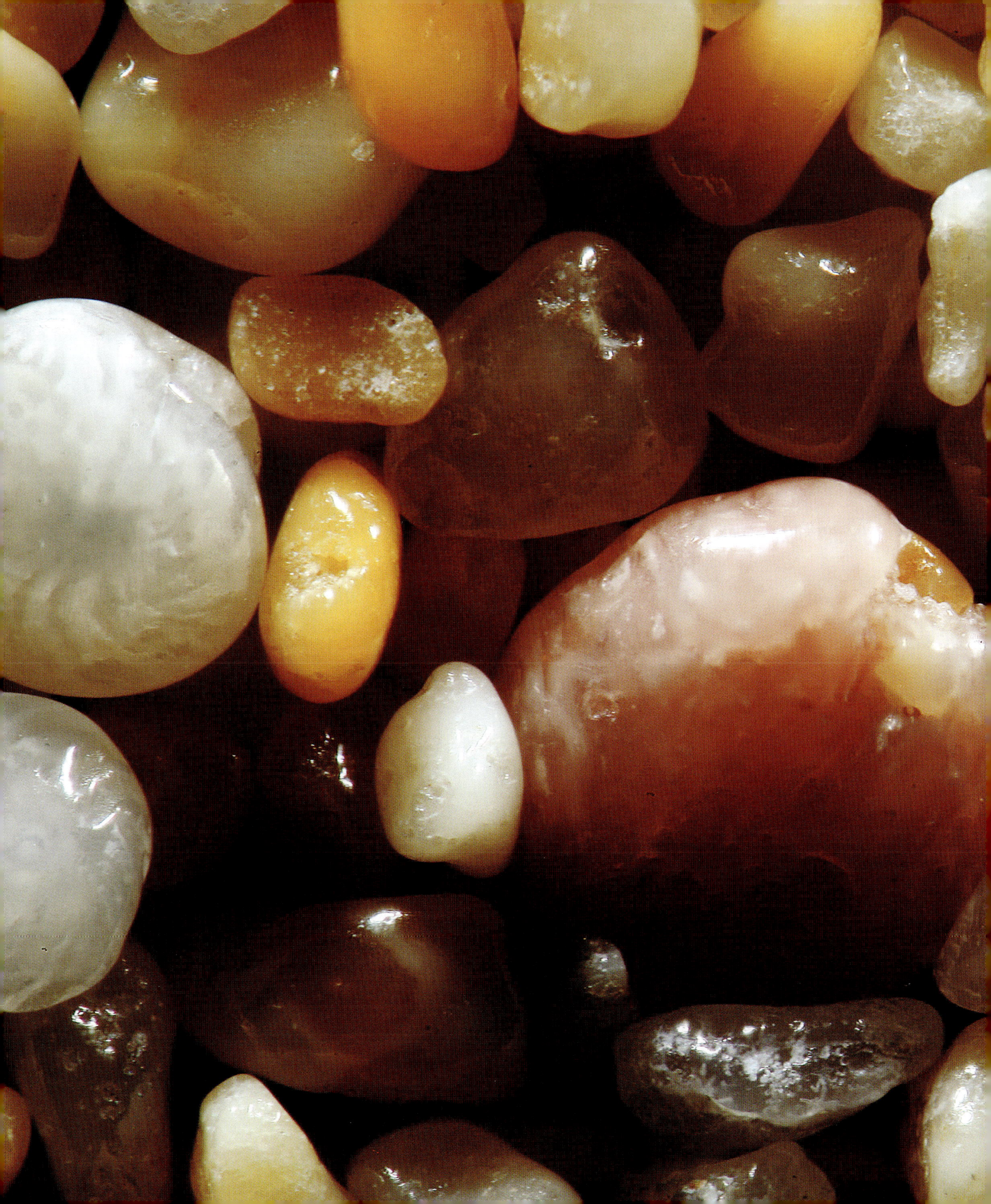

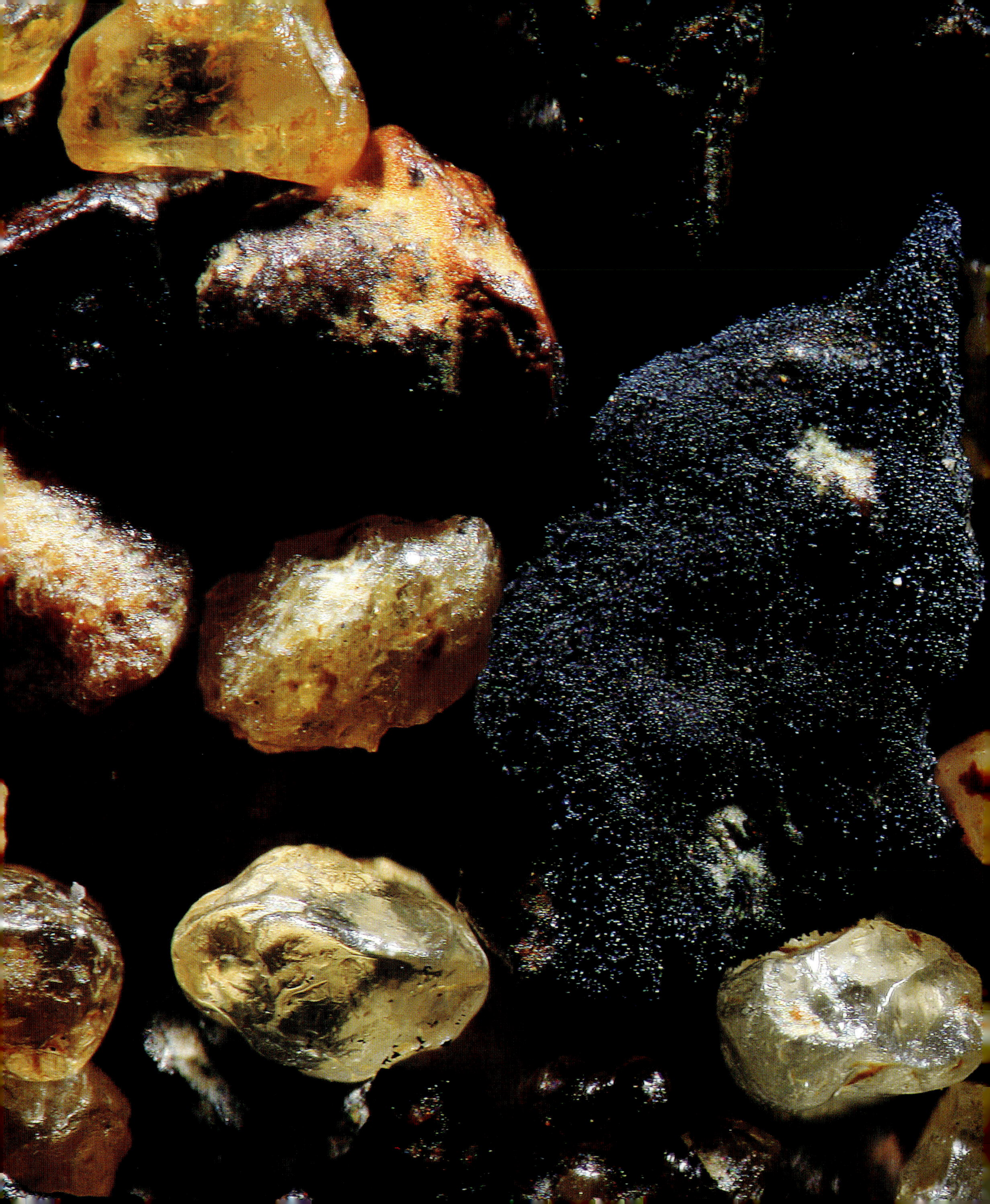

Dedication

I want to express my deepest gratitude by dedicating this book to my lifelong mentor, Professor Geoffrey Burnstock, for helping to guide me through the extraordinary world of science and nature and for teaching me the enormous value of creativity and originality.

To see a world in a grain of sand,
And a heaven in a wild flower.
Hold infinity in the palm of your hand,
And eternity in an hour.

—William Blake

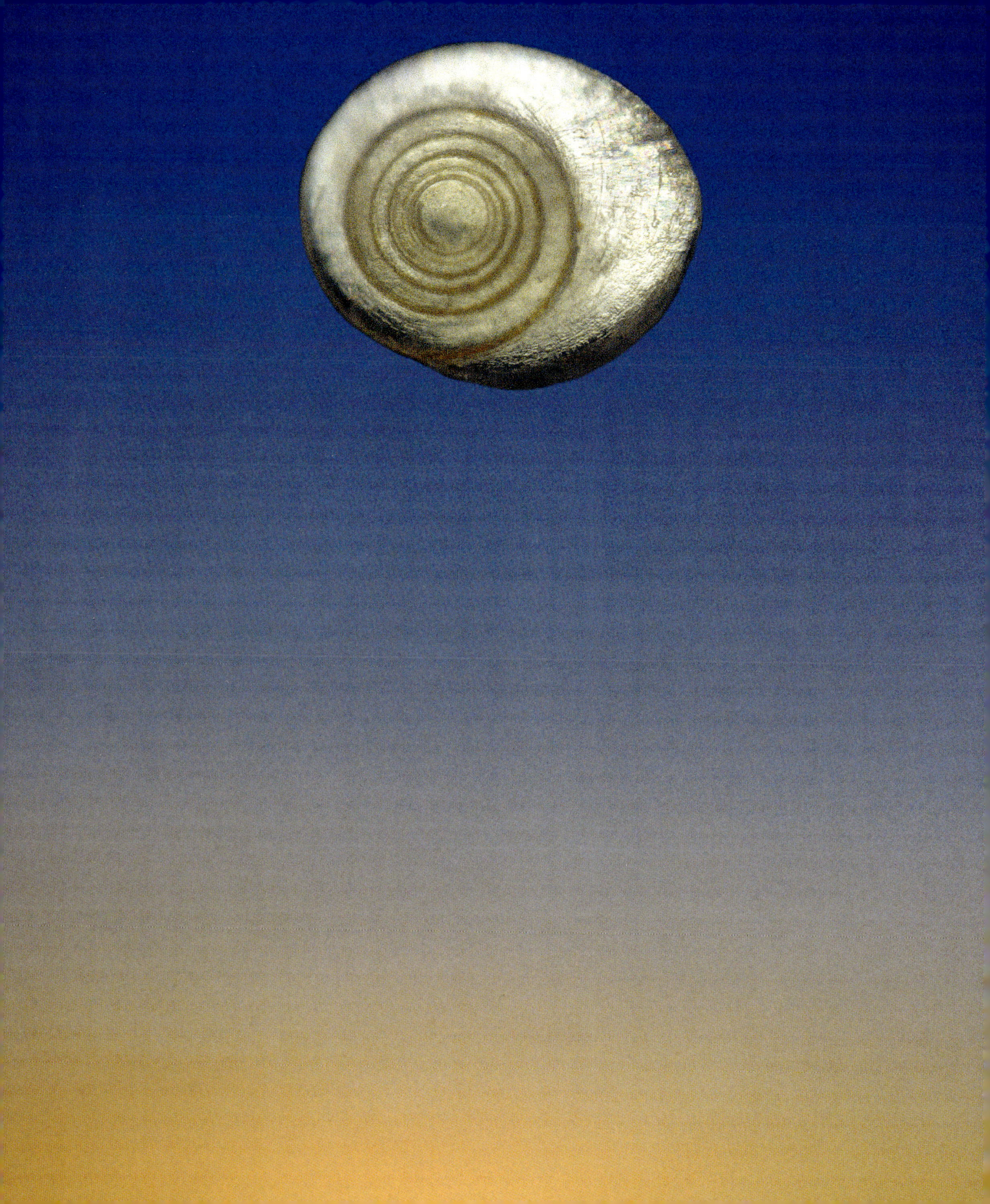

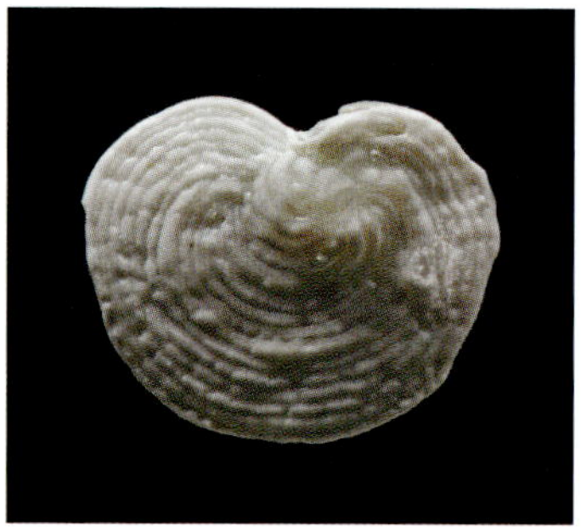
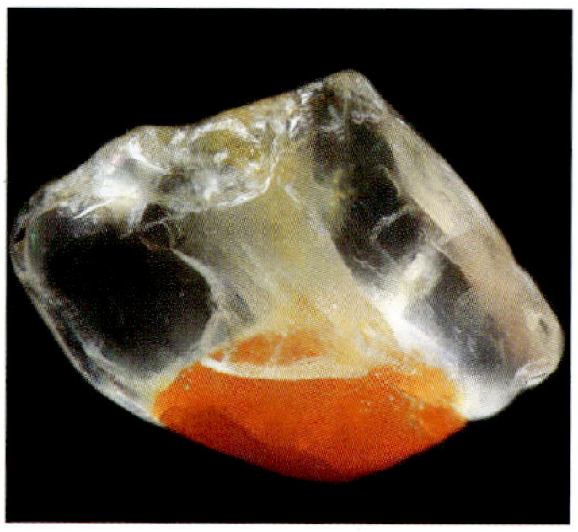
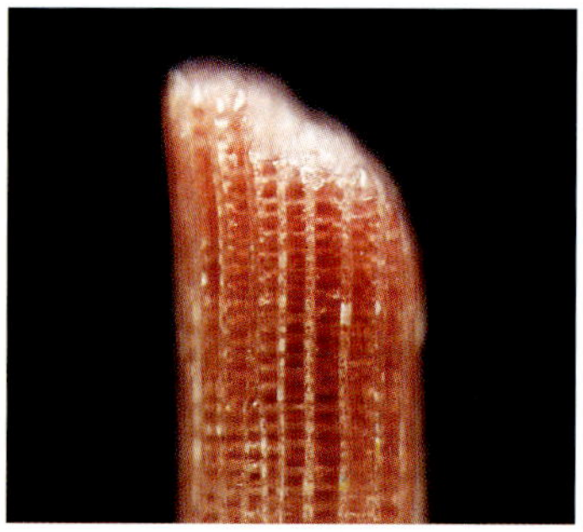

Contents

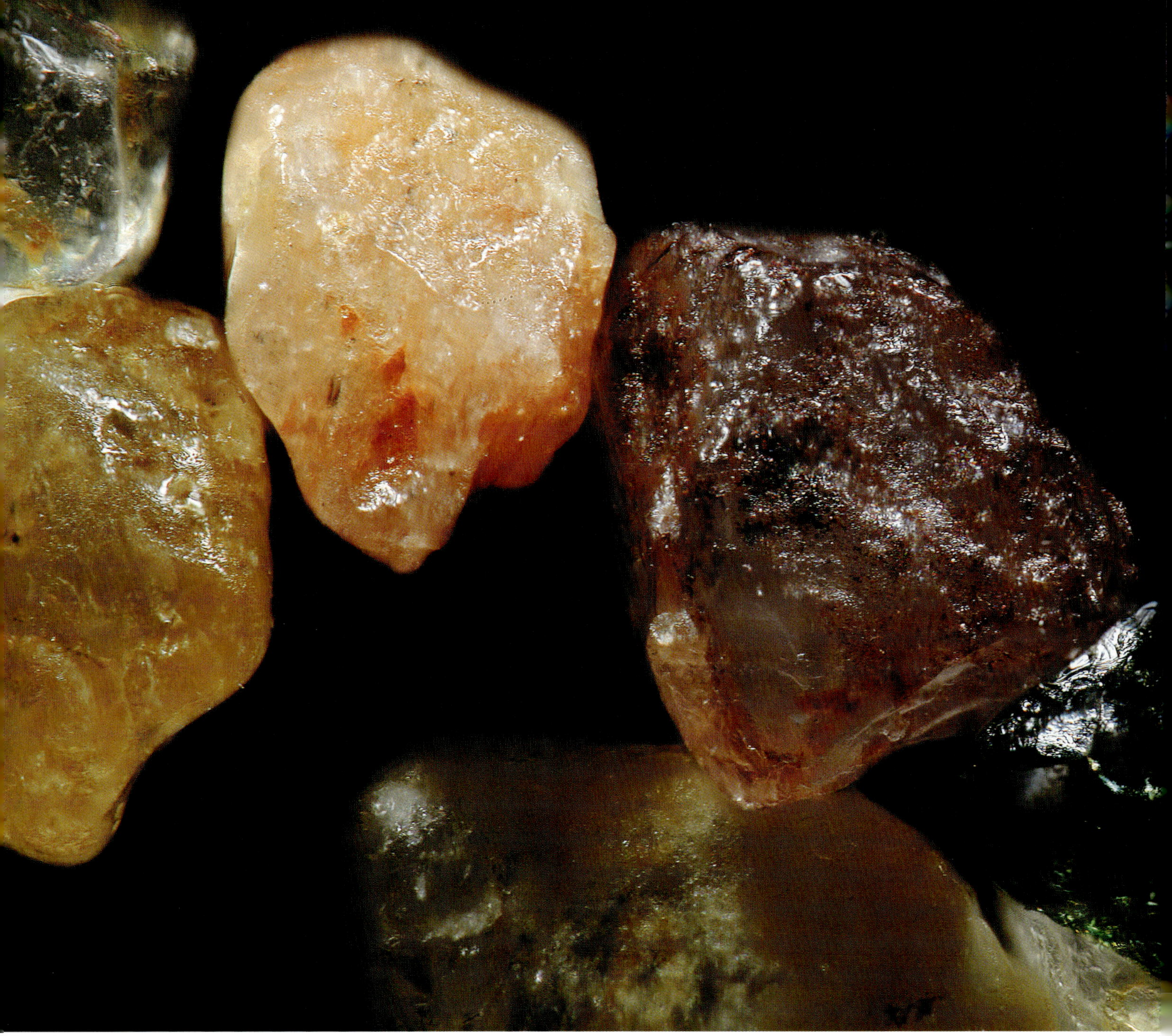

Acknowledgments

I wish to convey a special thanks to Shig Katayama, Bill Gasco, and Olivier Degremont, whose vision and support enabled the creation of the Edge 3D Microscopes, which were used to create this book.

I express my sincere thanks to Kate Hintz for her vast generosity, expertise, and support in the writing of this book.

Many others gave of their time and effort and provided beautifully collected sand samples from all around the world. I wish to convey my heartfelt thanks to Michael Adell, Michael Brucker, Dr. Aviva Burnstock, Tammy Burnstock, Cathy Carlson, Jim Christensen, Mickey Eskimo, David Greenberg, Stacy Keach, Leo Kenny, Martin Kerver, Deva Kranto, Fumihiko Kusama, Nancy LaRive, Daniel Lenington, Andrew Lumsden, Sara Martinez, Jean McDonald, Susan McDonald, Eoin McLaughlin, Ani Mestre, Robin and Bob Park, Rob Ratkowski, Jayne Rockmill, Kate Saunders, Asher Sela, Stuart Small, Still Point, Tom Talbot, Masako Tanaka, Professor Ed Tarvyd, George Van Noy, Kathleen Wets, and G. A. Wimer (Dane Photography).

I especially want to thank the creative people whose dedication and expertise turned a simple idea like a grain of sand into this beautiful book. I am extremely grateful to the publisher, Michael Dregni, who sparked this project into motion with his vision to create a book about sand grains. I am indebted to Danielle Ibister, the editor, who did an absolutely stellar job of shaping my words and ideas into the final presentation. I want to express my deep appreciation to Jayne Rockmill, my literary agent, for all her help and support during this project. And finally, thanks to Maria Friedrich, who created the beautiful book design.

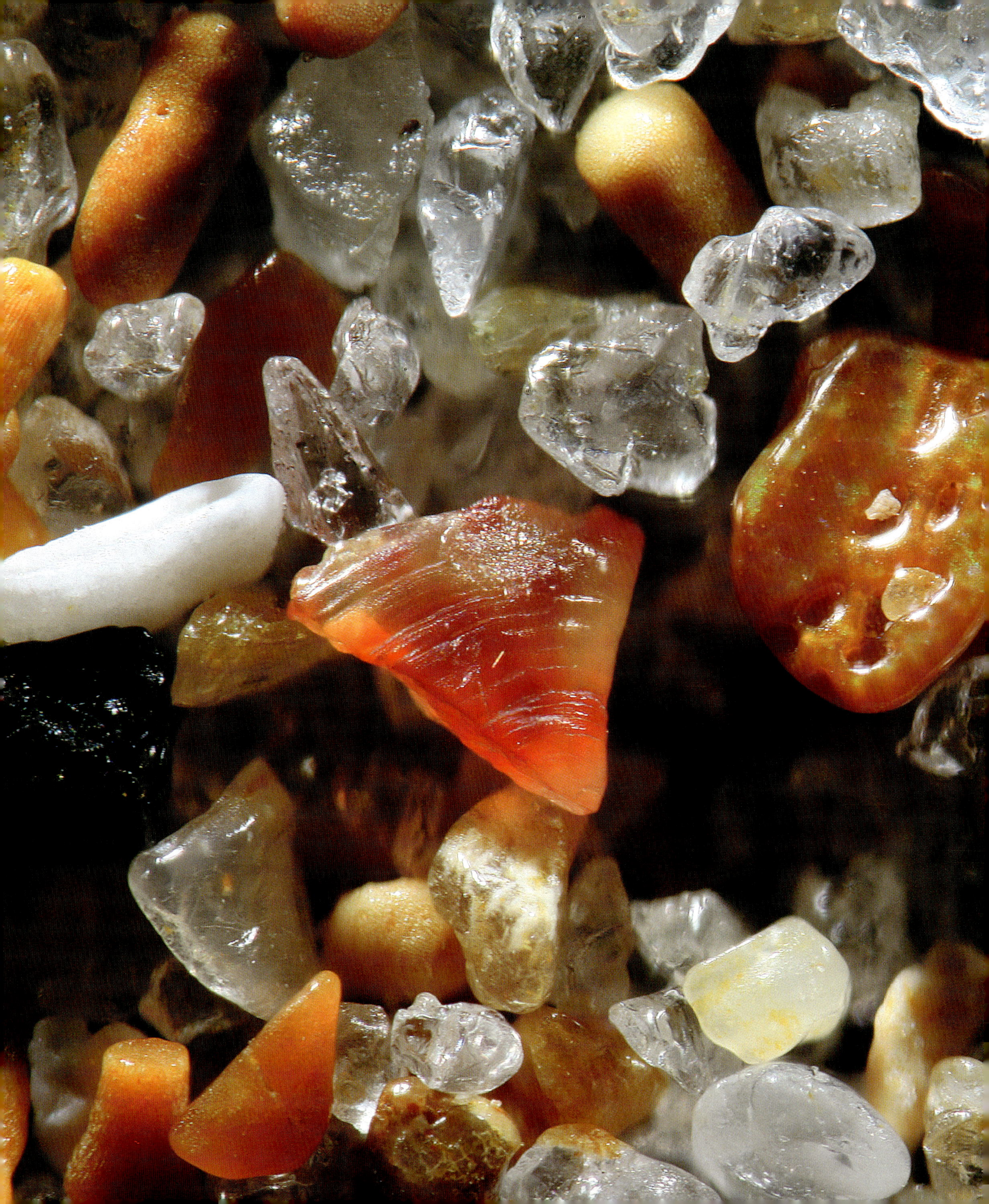

Foreword

By Stacy Keach

When I was a young boy, my parents held my hands as I walked on a sandy beach and encountered the waves beneath my feet for the first time. At first, it terrified me. The tide rushed back to join the sea under my toes, and I had the sensation of being swept under. Were it not for the fact that I held onto my parents' hands for dear life, I felt sure I was a goner. The fear subsided as I became aware that the waves were part of a natural process. My relief turned to joy, and I began to jump in the foam. Plunging my hands into the sand beneath my feet, I became mesmerized by the tiny dancing grains slipping through my fingers. So small, I thought to myself, and so many of them.

Some years later, I looked through a telescope at the stars for the first time. I remembered the claim that there are as many stars in the universe as all the grains of sand on all the beaches of the world. In that instant, I made the connection between these amazing tiny particles and the whole of the universe. Now, looking at Dr. Gary Greenberg's miraculous grains of sand photographed through a microscope, I am once again mesmerized by a whole new dimension. *A Grain of Sand: Nature's Secret Wonder* takes us to a captivating world that none of us has seen before.

Following in the tradition of the Renaissance, my good friend Gary—an artist, scientist, and inventor—is one of a growing number of specialists whose body of work reflects the synthesis of art and science. A prolific inventor with a number of patents for high-definition 3-D light microscopes, he has utilized his scientific skills with the microscope and the camera to create artistic landscapes of extraordinary beauty and fascinating detail. *A Grain of Sand* is a perfect example of that expression.

I first became acquainted with Gary's talents in the early 1970s when I joined the advisory board of Environmental Communications, a group of photographers and artists located in Venice, California, and devoted to documenting both the art and architecture of our cultural environment. It was a privilege for me to share in the expression of a movement where the worlds of art, science, and culture merged.

During this period, Gary also initiated a science club, where friends met once a week to discuss articles from *Scientific American*, *Science*, and *Nature* and ponder the implications of everything from black holes and string theory to the big bang and global warming. This period of my life would prove to be an invaluable inspiration for my work. I was just beginning to embark on an ancillary career as a narrator for scientific documentaries, especially the *NOVA* series for the Public Broadcasting Service. The knowledge I gained from the science club discussions not only deepened my understanding of complicated scientific principles, but it also enhanced my confidence to do my job as an informed authority. The discussions provided me with the opportunity to see and understand things about nature that had not been previously available to me.

Likewise, *A Grain of Sand* provides the reader with a delightful visual trip into a world beyond the reaches of everyday perceptions. It is an exploration into the wondrous landscape of things we do not see with our naked eye. It inspires the human spirit to know that the beauty and the mysteries of our universe are tangible and real, and that our lives are enriched by their presence. I am confident that you will enjoy the journey ahead.

Shell fragments and quartz sand grains from Geriba Beach, Brazil (magnification 300x).

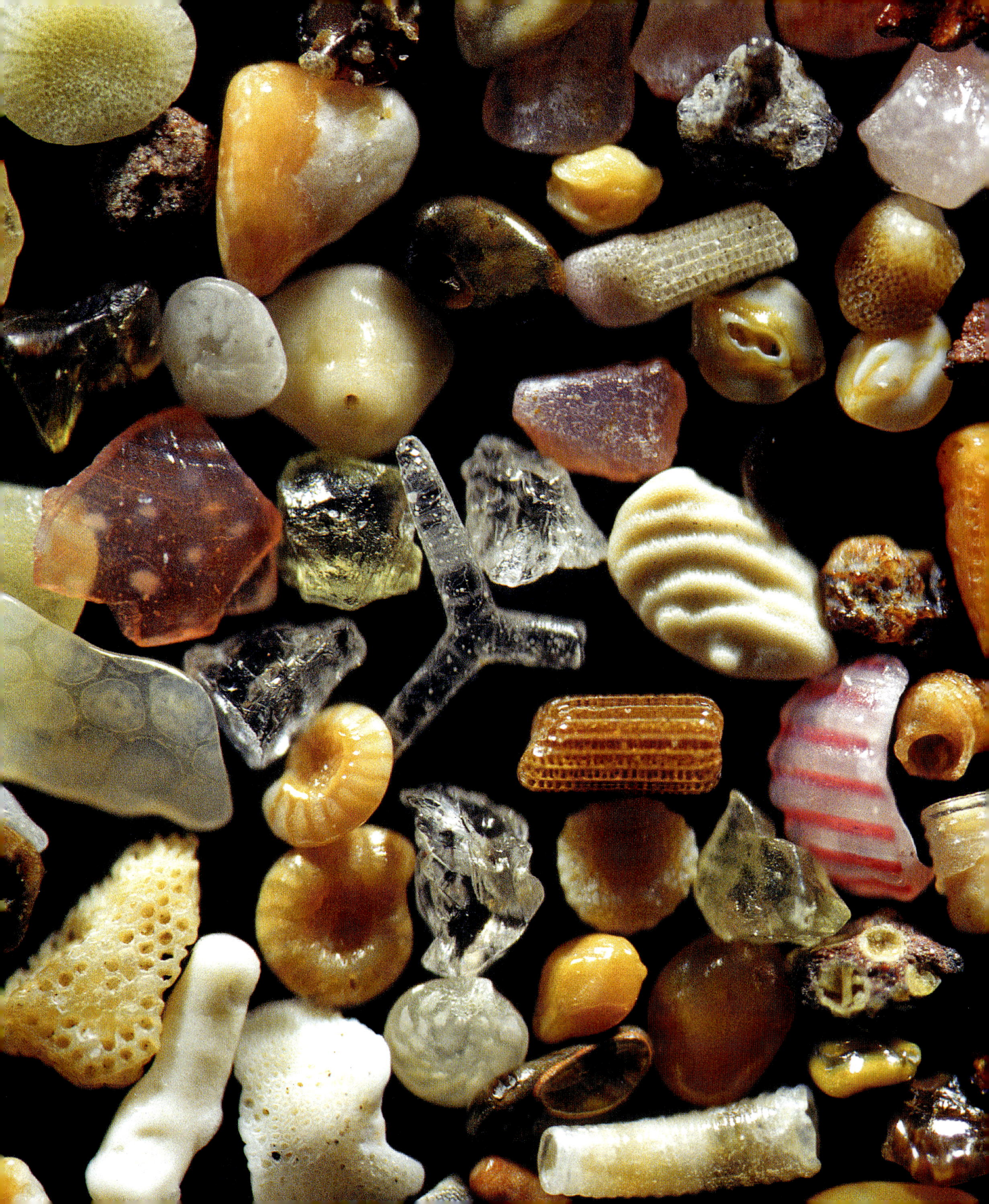

Introduction

People often ask me, "How in the world did you get interested in looking at grains of sand through the microscope?" Half kidding, I tell them I had a lot time on my hands. Actually, the story began in the year 2000, when my brother David sent me a film can full of sand from Maui as a subtle hint to get me to visit him there. The sand sat on my shelf for months, until one day in 2001, when the film can labeled "Maui Sand" caught my eye. I put the sand under the microscope, wondering what it looked like. I expected to see a bunch of tiny brown rocks. But, wow, was I surprised!

This is what I saw: spectacular colors, shapes, and textures and all kinds of wonderful remnants of volcanic rock mixed with bits of dead biological stuff, very beautiful to behold. I felt like a kid who just discovered hidden treasure.

Wondering what other beaches would look like through the microscope, I contacted an old friend who lives in the Virgin Islands, and he sent me a couple film cans full of sand. I was not disappointed. One was labeled "Flamenco Beach, Spanish Virgin Islands." It contained fine grains of sand chock-full of fabulous bits of coral in exquisite colors and strewn here and there with tiny, microscopic shells.

I have collected and examined sand from all over the world ever since. I never lose my curiosity when opening a package someone has sent me from a gorgeous, remote beach in the Pacific Islands, the South China Sea, or some other extraordinary place. The astonishing thing is that, when viewed through the microscope, sand grains from every beach are very different. In fact, no two grains of sand are exactly alike.

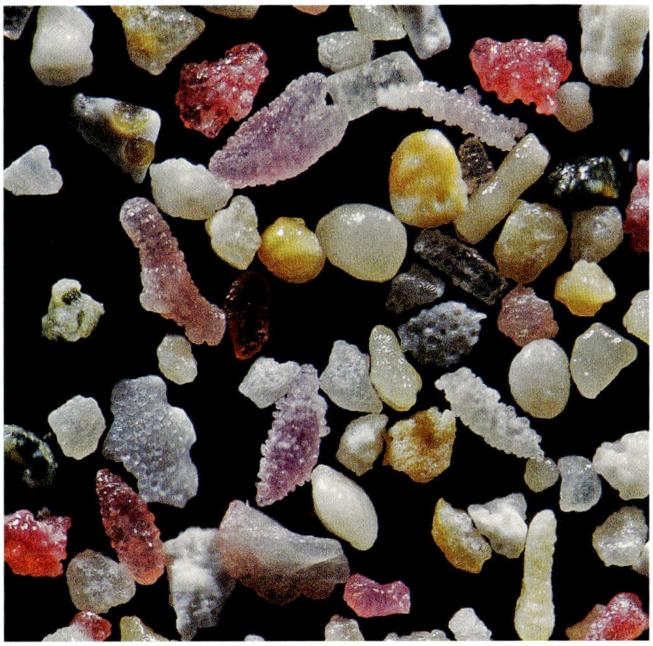

The second sample of sand that I examined, from Flamenco Beach, consisted of very fine grains made mostly of brightly colored bits of coral, forams, broken bits of sea urchin spines, and bits of silica and calcium carbonate (magnification 35x).

These grains of sand from Maui were the first that I examined through the microscope. I was amazed by the array of spectacular bits and pieces. The Y-shaped glassy structure is a sponge spicule, which functions as the internal skeleton of sponges. To the left and down are two perfectly formed microscopic shells. To the right and down from the sponge spicule is a bit of brown sea urchin spine showing its intricate structural design, and to the right of that is a pink-and-white bit of a broken seashell. At the very bottom is a tiny white tube-building worm (magnification 120x).

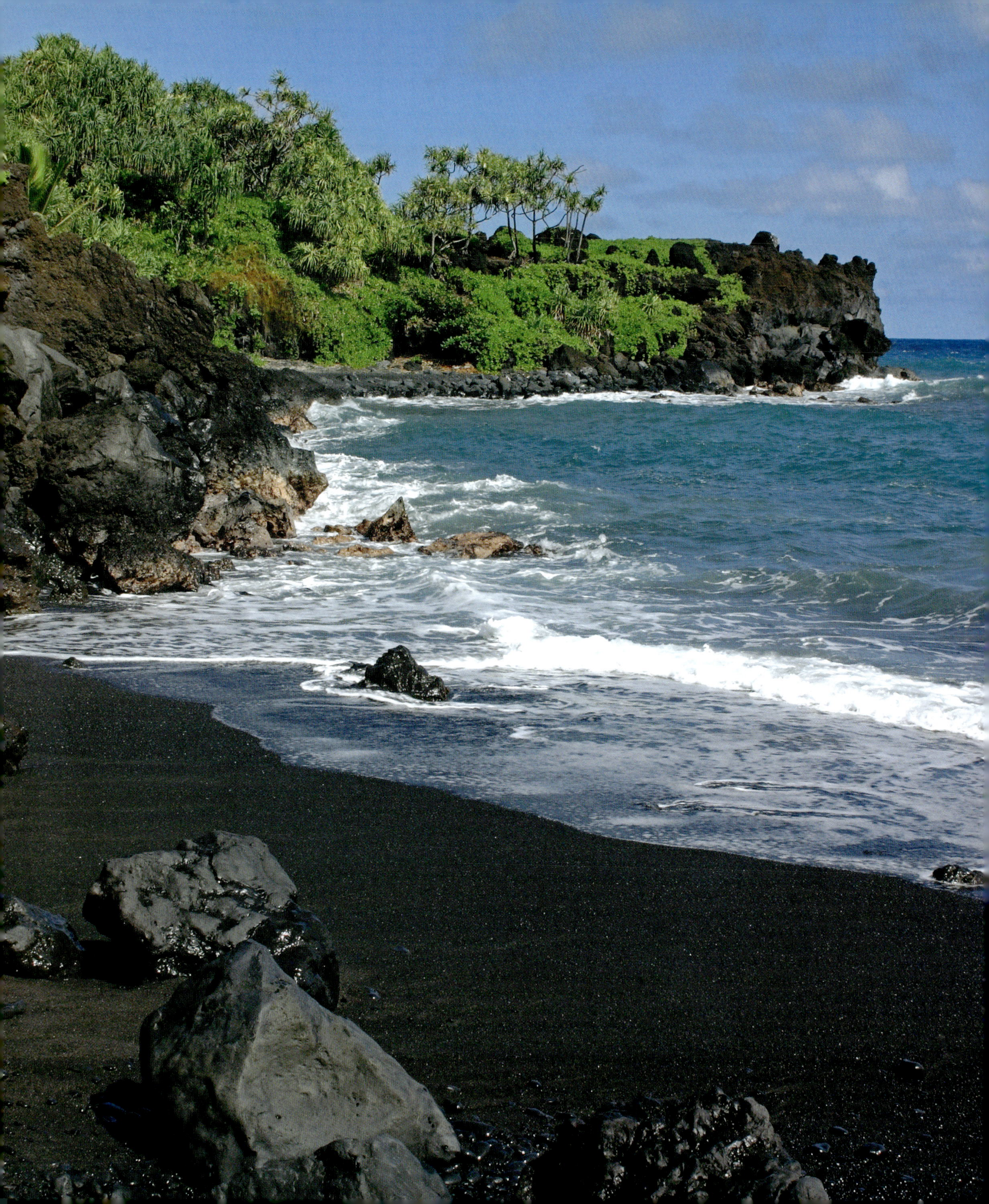

CHAPTER 1

A Signature Written in Sand

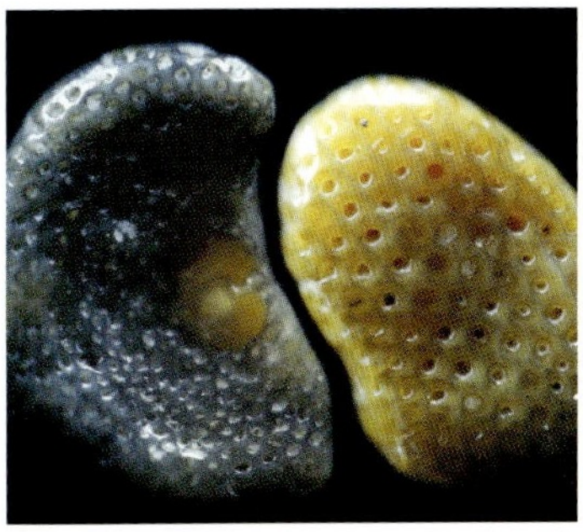
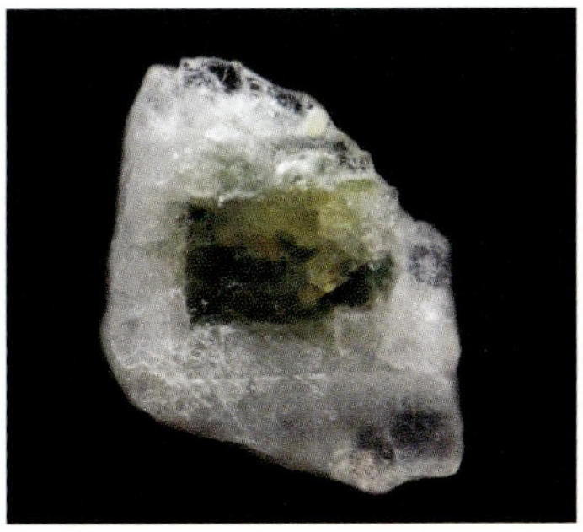
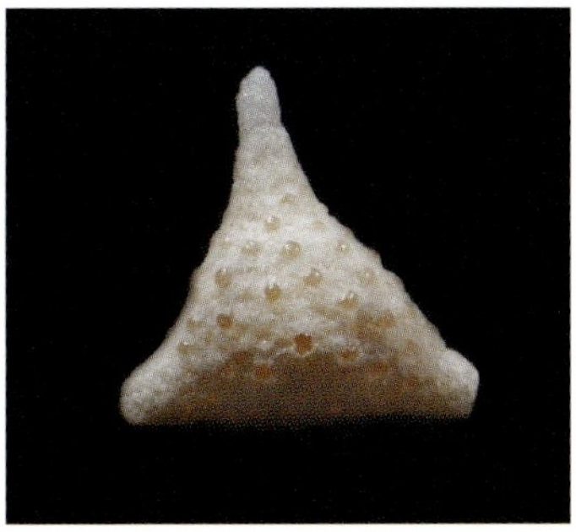

The composition of the world's beaches is unexpectedly diverse. Consider the beaches of the Hawaiian Islands, ranging from the beautiful green-sand beach at South Point on the Big Island, composed of pure olivine crystals; to the red-sand beaches of Maui; the many lunarlike, black-sand beaches; and the soft, rich yellow of Big Beach at Makena, Maui. The character of sand can vary tremendously from beach to beach, even on beaches that are virtually adjacent to one another. Right next to the rich, yellow sand on Makena Big Beach is the red beach of Makena Point, and just beyond that is Makena Bay, which looks completely different from either Makena Point or Makena Big Beach. Yet, all these beaches are little more than a mile apart.

Because of their volcanic origins, Hawaiian sand grains contain fabulous multicolored minerals and volcanic rocks that have surprisingly beautiful sculptural shapes. Also, Hawaii teems with ocean life, so the sand is sprinkled with little bits of broken seashells, corals, sea urchin spines, sponge spicules, and the intricate microscopic houses of tiny single-celled protozoa. The huge variety of sand grains is astounding, and each one has a story to tell.

Sand grains eroded from volcanic rock create Maui's black sand beaches.
Michael J. Thompson, Shutterstock

Examining sand grains through the microscope is a wonderful way to find out about the biology, geology, and ecology of the local environment. Sand holds many clues to the makeup of the rocks, mountains, and soil, and it provides a biological record of the history of marine life. When we walk along a beach, we unknowingly tread upon millions of years of biological and geological history.

Each beach has a signature, so to speak, that is written in sand. These messages have provided critical information to forensic scientists in life-or-death situations. An interesting example comes from World War II. In 1944, thousands of balloons carrying bombs were landing in Wyoming and Montana. The balloons were made of rice paper, with a basket suspended below carrying a bomb encircled by a few dozen sandbags. An altimeter was attached to the apparatus, part of an elaborate device to maintain altitude day and night on the journey from its place of origin. The sandbags were part of the control system; when a sandbag was released, the balloon gained altitude. At first, the government assumed the balloons came from Oregon or Washington, but the sand in the sandbags didn't match any beach in the Northwest United States. Presumably, it came from somewhere in Japan.

The question was: where in Japan were the balloon bombs being made? And how could they survive such a long journey? It turns out that Japanese scientists had discovered a weather pattern that we know today as the jet stream. They realized they could sail a balloon from Japan to the United States in about thirty-three hours if they sent it high into the atmosphere.

American scientists figured that if they could pinpoint the origin of the sand that filled the sandbags, they could locate the factory where the bombs were being made. They examined the sand grains using high-power microscopes. Three important clues surfaced immediately. First, there were no

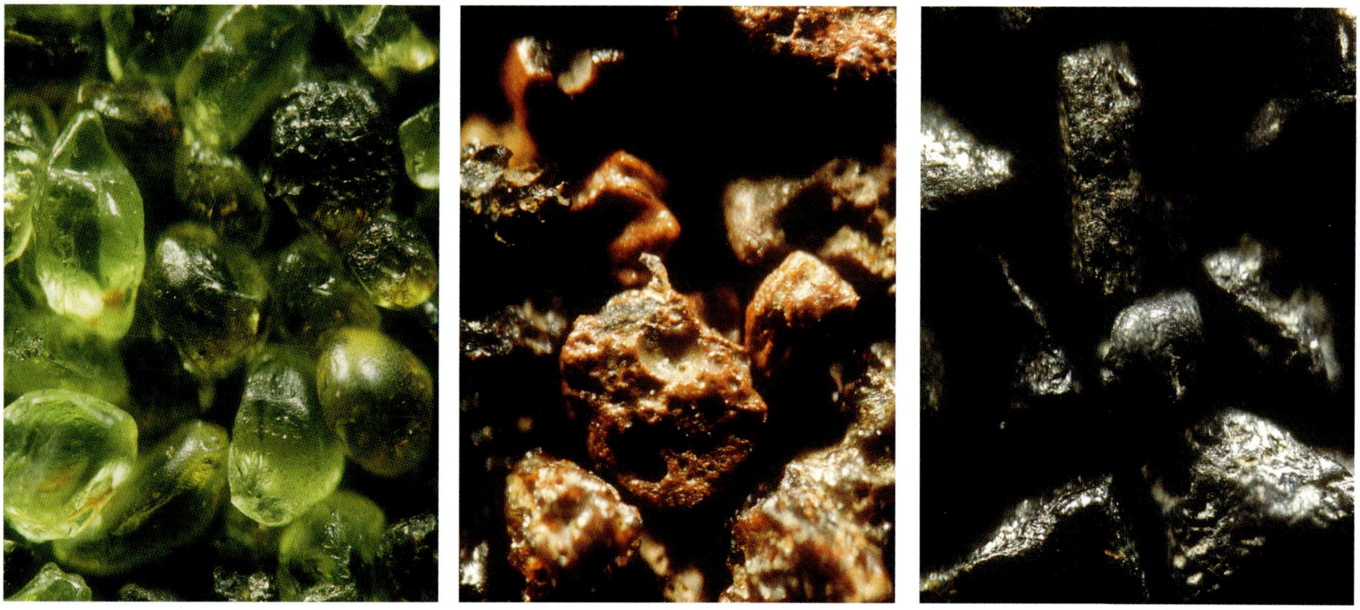

The sand on Hawaii's beaches varies in composition and color. *Left*: The green sand on South Point on the Big Island is made of pure olivine crystals. *Center*: The red sand grains on Hana, Maui, are volcanic in origin. Iron oxide makes them red. *Right*: Black sand beaches are common in Hawaii. The black volcanic sand grains are from Hana (magnification 70x).

granite or quartz crystal sand grains present, which would be characteristic of sand from continental landmasses. This clue made it likely that the sand was from an island. Second, there was no coral in the sand. That meant the sand came from north of the 35th parallel, because bits of coral are normally found in sand from tropical zones. And third, the sand contained a large amount of volcanic rock made up of rare minerals.

While geologists tracked down which areas of Japan contained such minerals, biologists studied the microscopic fossils found in the sand. From the mineral evidence, it looked like the sand came from somewhere close to Tokyo. But where? When the forensic scientists looked with the electron microscope, they discovered a rare type of diatom amongst the grains of sand. Diatoms are algae that have intricate silicate shells. They are one of the main components of plankton, and each species of diatom constructs a different characteristic, microscopic shell. Scientists searched the libraries to find out where this particular diatom lives and finally came across the notes of a French expedition that discovered that particular diatom at a site in Japan in 1889.

With the suspected location in hand, reconnaissance flights verified the exact site of the factory. Nine thousand balloon bombs landed in the United States from 1944 to early 1945. Many of them exploded harmlessly in the middle of the great Northwest. After U.S. bombers destroyed the factory, no more balloon bombs were seen.

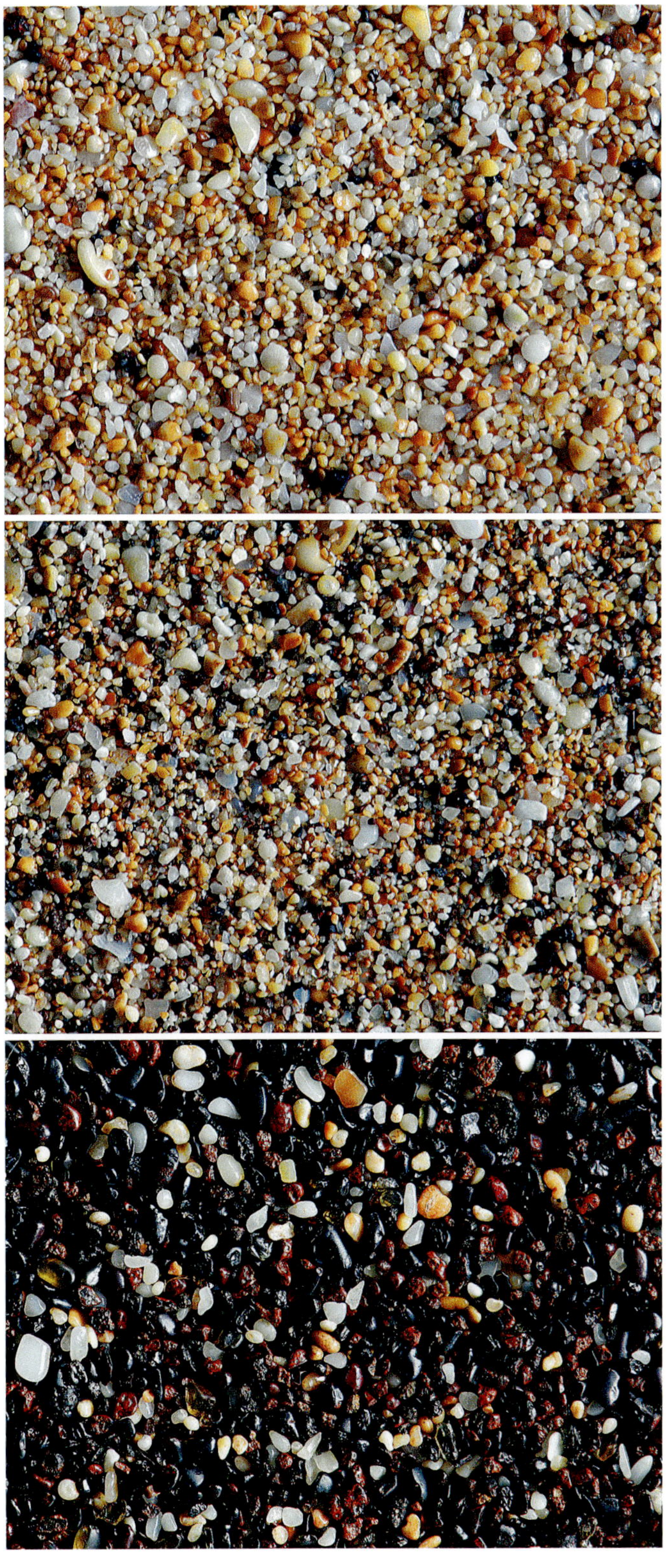

Sand grains from neighboring beaches can look quite different, as evidenced by the sand from (*top*) Makena Big Beach, (*middle*) Makena Bay, and (*bottom*) Makena Point. The three beaches are little more than a mile apart from one another (magnification 3x).

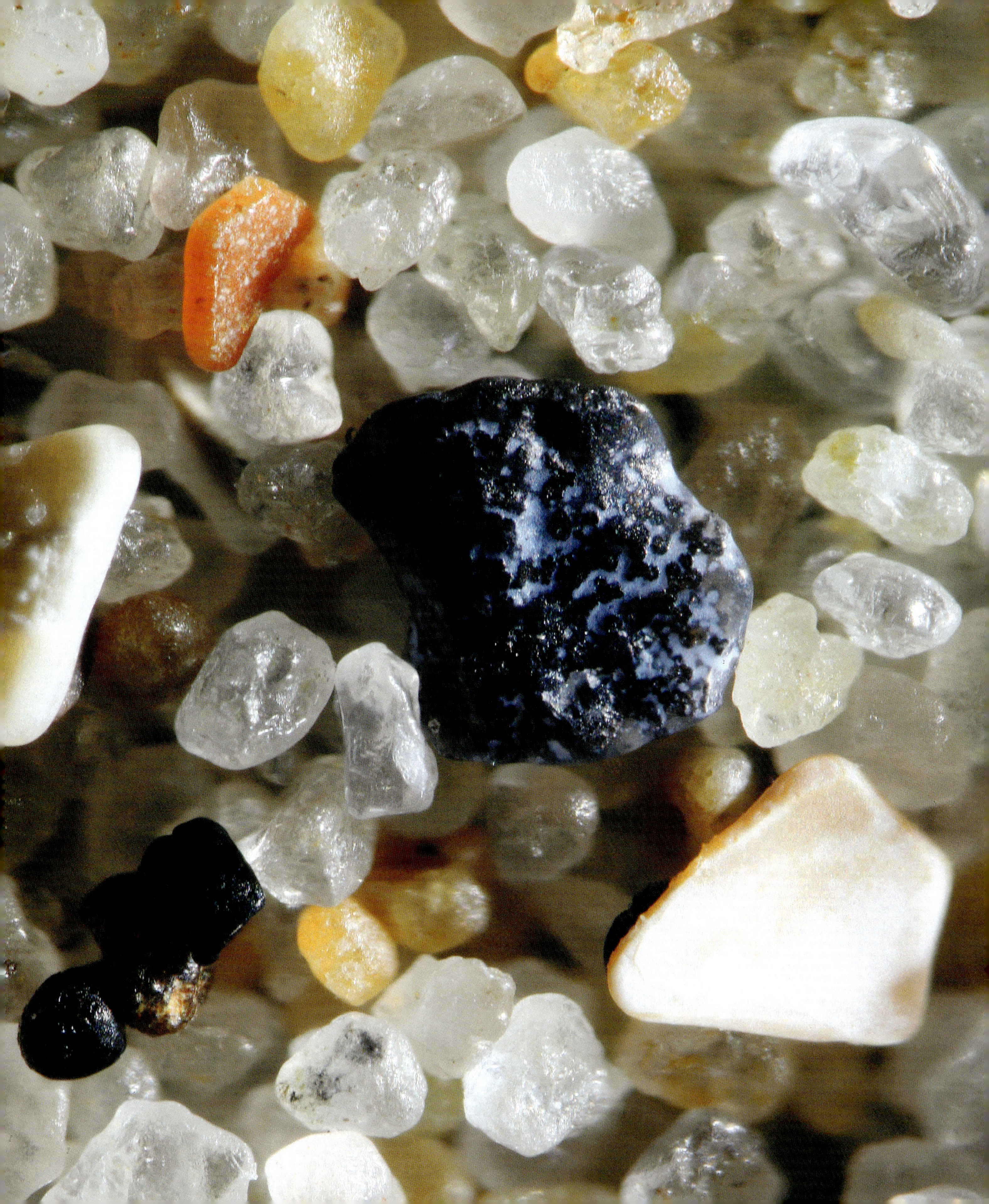

Chapter 2

Erasing the Line between Science and Art

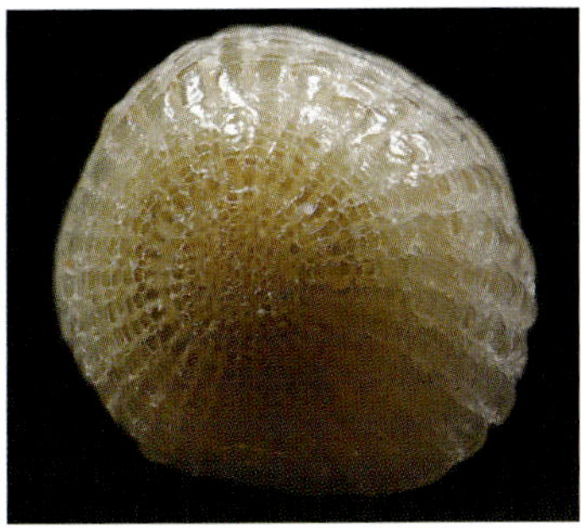
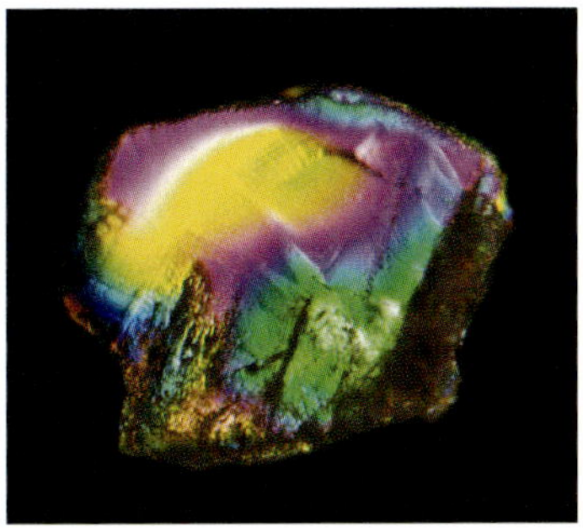
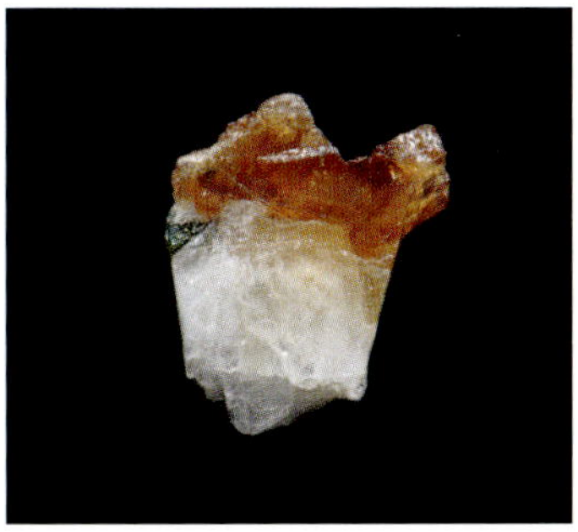

The mysteries of science and nature intrigued me at an early age. I recall being in junior high school and my father returning home from a trip to Japan with two microscopes: one for our family physician and one for me. I was thrilled! Mine was a good stereo dissecting microscope. I explored the neighborhood for cool things to look at through my microscope. I was amazed by the beauty and complexity of ordinary things, such as bugs and flowers and rocks. My microscope transported me into a new dimension, where I was no longer limited by what I could see with my naked eye.

This experience made me recognize that our perceptions are limited by our sensory abilities. Our consciousness is locked into a fixed point of view in time and space. For me, the microscope was like a pair of magic glasses that revealed the underlying nature of reality. I felt like a kid with a pair of genuine x-ray glasses.

I became interested in photography at the University of California, where I majored in psychology. When I graduated, my grandpa asked me what I planned to do with my life. I told him I wanted to be a photographer, prepared for him to ask me, "What about the four years you just spent studying psychology?"

To my surprise, he said, "You know, Gary, I bet you're going to show people things they've never seen before."

A piece of industrial slag becomes a black-and-blue sand grain on a quartz beach in Belgium (magnification 125x).

With my camera at my side, the entire world became an adventure. Photography helped me realize that the universe is continuously unfolding—always new, always different. Inspired by this notion, I became a photographer and filmmaker and began to develop a personal vision of the world around me.

I had the very good fortune to find a brilliant mentor early in my life. At the age of twenty, I had taken a hiatus from university to spend some time in Australia, where I became good friends with British transplant Geoff Burnstock. At the time, Geoff was thirty-five years old and had just been appointed full professor and head of the Department of Zoology at the University of Melbourne.

Over the years, Geoff occasionally hired me to make 16mm time-lapse films of living cells through the microscope. I was fascinated by what I saw: the true fabric of life before my very eyes. In 1975, Professor Burnstock returned to the United Kingdom and became the chairman of the Department of Anatomy and Developmental Biology at University College London. The following year, he asked me to join his new group as a graduate student. Not wanting to miss this extraordinary opportunity, I moved from Los Angeles to London at the age of thirty-three in order to earn a Ph.D. in biomedical research.

Later, as an assistant professor at the University of Southern California, I employed microscopes as a tool for investigating the formation of birth defects. Along with Professor Michael Melnick, I studied the development of the neural tube in chick embryos in order to better understand the dynamics of human birth defects such as spina bifida.

Even though the microscopes I was using were the best money could buy, they didn't allow me to see the degree of detail I needed. Microscopes are good at producing sharp images, but only if the specimens are sliced into very thin sections. The average thickness of specimens prepared for a high-power light microscope is between 1 and 5 micrometers, whereas the diameter of a single cell is between 10 and 20 micrometers (1 micrometer = 1/1000 of a millimeter, or a millionth of a meter). So you normally don't see an entire cell because it's larger than the slice being viewed. Our problem was that we needed to examine a whole neural tube, which is more than 100 micrometers thick. We couldn't slice it into sections because we needed to observe it alive. We needed to study its dynamics using time-lapse video analysis in order to reveal how the neural tube develops over time.

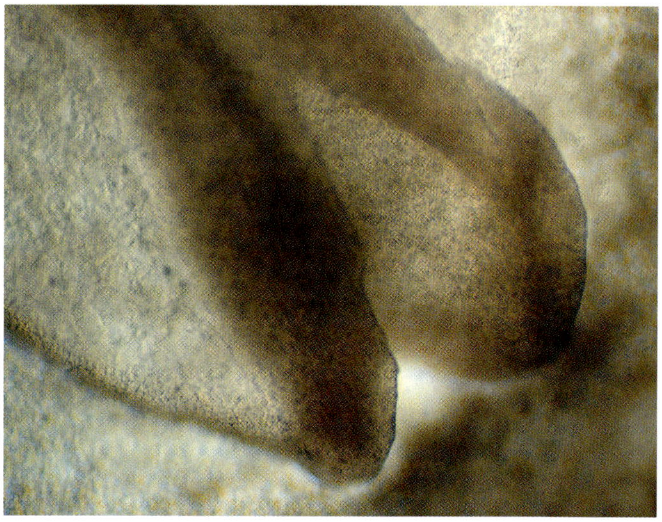

The beginning of neural tube formation can be seen in a living chick embryo. Studying such specimens compelled me to develop a microscope lighting system to better visualize transparent living tissues rather than thinly sliced tissues that have been fixed and stained (magnification 125x).

I set out to improve the resolution, contrast, and depth of focus of my images. One of the problems was that high-power lenses have a shallow depth of focus. In other words, only a thin optical slice is in sharp focus, and both the foreground and background are out of focus. Another problem was the lack of contrast, because a thick specimen such as an embryo scatters a lot of light, making everything appear as though you're looking through a veil.

I suspected that the solution might lie in the microscope's illumination system. I experimented with different types of lighting conditions. I rigged an apparatus that allowed me to illuminate the specimen from extremely oblique angles, so that the light source was no longer directed straight into my eyes. The result was dramatic. Instead of the light being scattered into the eyepiece, the light was scattered off to the side. This new lighting arrangement was just what I needed. The oblique lighting created much sharper contrast in the embryo photos, and it dramatically improved the depth of focus. It also provided about 40 percent better resolution compared to a conventional microscope illumination system.

A couple more years of experimentation led to my first patented invention, which was a unique form of lighting for microscopes that utilizes several independent oblique light sources simultaneously.

Peripheral vision and lateral thinking are critical to the process of invention. The lessons I had learned about lighting as a young photographer I later applied to microscope illumination systems. Normally, microscopes use a single light source, either passing through the specimen or reflecting off of it. However, as a photographer, I knew that lighting a scene with one light source pointing straight at the subject is not the best way to bring out details in the picture, because the lighting is too flat. Every good photographer knows that the artful use of lighting is essential to creating a dramatic and informative image.

One day, with great enthusiasm, I realized that my microscope illumination system might be a way of creating 3-D images in a standard high-power light microscope. Further experimenting with different forms of illumination, I noticed that oblique lighting had the strange effect of making the specimen appear tilted, as though you were viewing it from an angle. The more oblique the angle of illumination, the greater the apparent tilt in the specimen. I reasoned that if I used one oblique light from the left and another oblique light from the right, it would be possible to produce a left-eye and right-eye view of the sample simultaneously, which would result in a dramatic 3-D image. In other words, both eyes would receive a 2-D image but from a slightly different angle; the human brain then creates the third dimension. With the help of my mentor and

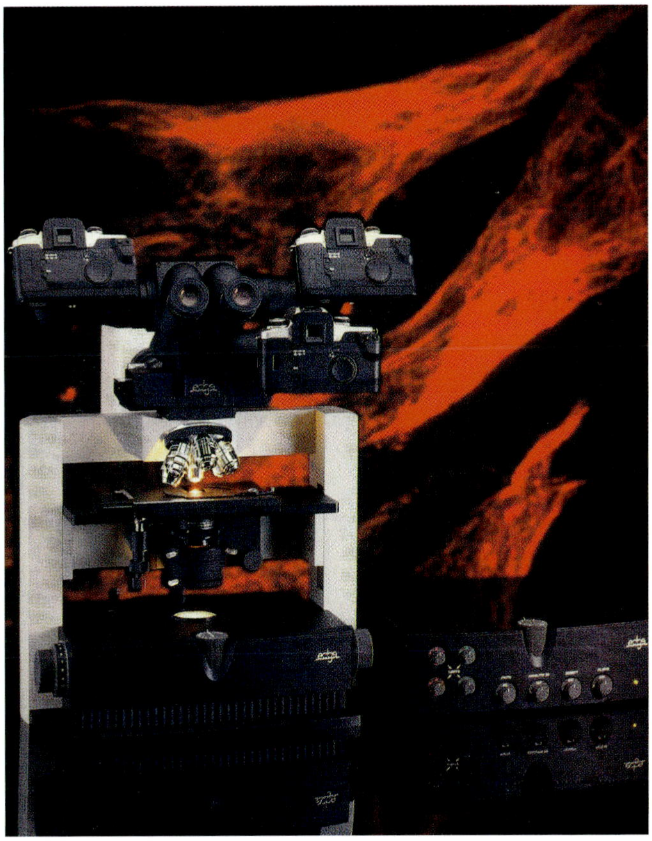

The Edge 3D Microscope was used to create the images in this book.

good friend Professor Alan Boyde at University College London, I built a working prototype in 1990, and, to our delight, it performed well enough for us to proceed with further development.

Low-power 3-D microscopes have been available for more than a hundred years, and they can achieve magnifications up to about 50 times. But no high-power 3-D microscopes were available that produced real-time stereo images of the specimen in the magnification range of 100 to 1,500 times. I became so captivated with the idea of developing a 3-D microscope that I resigned from USC in 1991 and cofounded Edge Scientific Instrument Corporation. We developed what was then the only commercially available, high-definition, real-time 3-D light microscope and sold it to research scientists throughout the world. I have been issued seventeen patents for 3-D microscopes, many of which have been manufactured and are currently being put to good use in research labs and universities. Today, the elite, cutting-edge microscopes are laser-illuminated, computer-controlled, robotically driven, and have sophisticated 3-D capabilities.

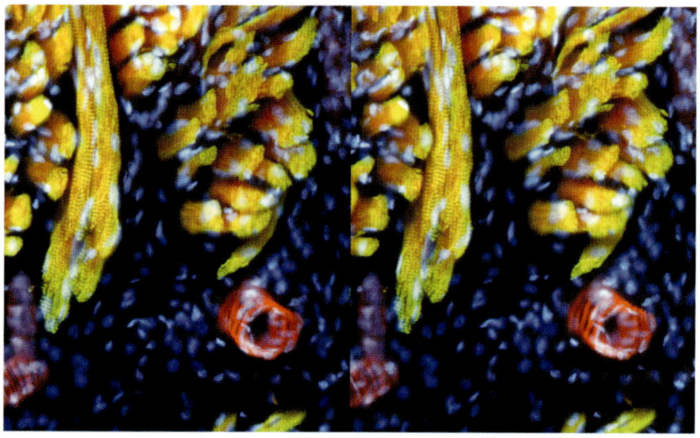

Three dimensions of information are tracked in this image of a mouse tongue. The red color labels a protein specific to smooth muscle cells, thus the red tubular shape reveals the presence of a tiny arteriole. The yellow striped fibers are muscle cells. The yellow color is created from the combination of two proteins, actin and myosin, found specifically in muscle cells. The blue color marks the locations where DNA is present, disclosing the position of the nucleus of each cell. If you cross your eyes, the two images will merge into a single 3-D image (magnification 150x).

For three decades, I've been using microscopes to explore micro space in much the same way that telescopes explore outer space. I think of myself as a landscape artist creating images of unimaginable worlds—worlds where reality is seen as abstract form, color, motion, and texture.

My passion is to erase the line between science and art. Both are valuable ways to explore nature. Both reinforce one another. Art has always depended upon science and technology; more than ever, science depends upon art to bring information to the people. Art will lead the way in conveying science to the consciousness of twenty-first-century culture.

Since 2001, I have turned my attention to creating art through the microscope, rather than using microscopes only for scientific explorations. As I've examined bits and pieces of everyday things—pieces of food, flowers, and grains of sand—I've discovered an extraordinary universe in things we take for granted. The amazing thing is that almost everything is either beautiful or intriguing when looked at through the microscope. In 2003, I launched the website www.sandgrains.com to communicate my findings.

In order to create the most dramatic images of sand grains, it is important to collect the samples in a special way. Most people collect sand from the dry area of the beach, but the grains from the upper beach can look dull and pitted through the microscope because they have been eroded by the wind. In comparison, sand from the wet portion of the beach has been tumbled by the action of the surf, making the individual grains appear naturally shiny and bright.

I normally use fine-grained sand for my images. There is a severe problem with depth of focus when photographing through a microscope. The part of the image in sharp focus is just a few micrometers thick, so structures in the foreground and background are blurred. The higher the magnification, the thinner the in-focus slice, and both the background and foreground become more out-of-focus.

We have overcome this problem of shallow depth of focus in an interesting way. I utilize a sophisticated computer program employed in modern biomedical imaging laboratories. It works like this: First a photograph is taken that is focused at the top of the sand grain, then the microscope is slightly refocused and another image is recorded, and so on, until there is a stack of photographs, each one of which contains a different portion of the sand grain in sharp focus. The computer program analyzes each image in the stack and selects only the in-focus parts, discarding the out-of-focus parts. The in-focus portions are then seamlessly joined into a single image that is totally in-focus.

The magnifications that I generally use for sand grains are on the low end of the range for a high-power light microscope. I usually magnify the images anywhere from forty to four hundred times their actual size, whereas the microscope can magnify up to a couple thousand times. If I were to use these very high magnifications, you would only see a small part of the surface of a single grain. It would be like taking pictures of a person by focusing close-up on the pores rather than the face.

I strive to bring out lots of detail in the sand grains, which means focusing relatively closely, but at the same time I also like to portray the surroundings. Therefore, I occasionally create large panoramic images that are printed eight or ten feet long. Sometimes I select the sand grains by hand and arrange them on a glass slide before capturing the images. The process begins by carefully looking at a sample of sand through the microscope. The sand is spread out flat on a glass microscope slide, just one layer thick, so that each individual grain can be seen, one by one. Each grain is beautiful, but about one in every five hundred grains is spectacular. When I find

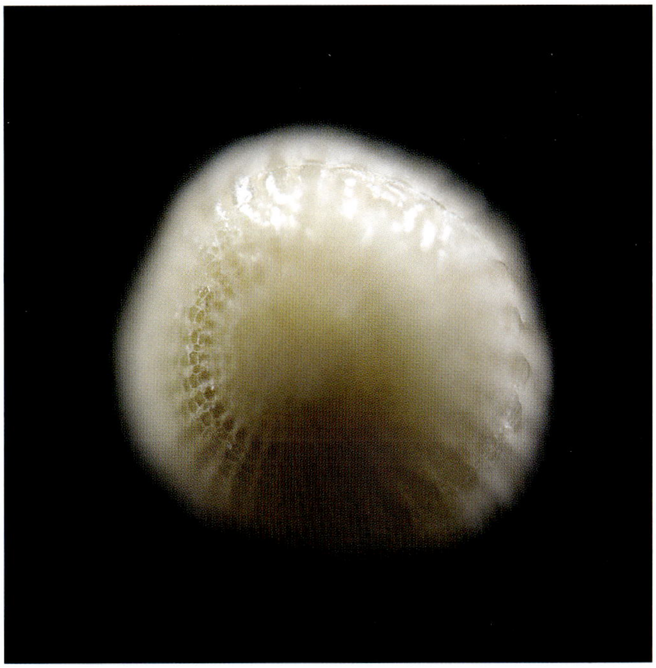

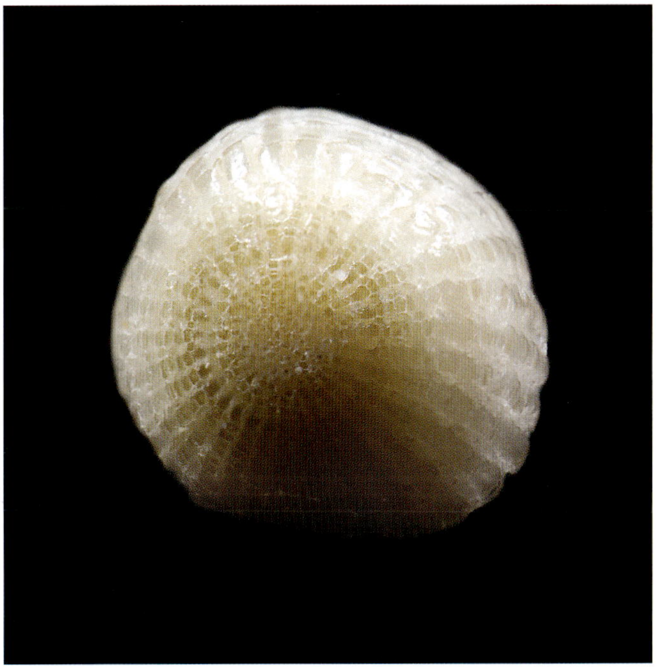

Top: A single image taken through the middle of a grain of sand shows a very shallow depth of focus. *Bottom*: An extended depth of focus picture is constructed from ten images of the same sand grain, each focused at a different level (magnification 110x).

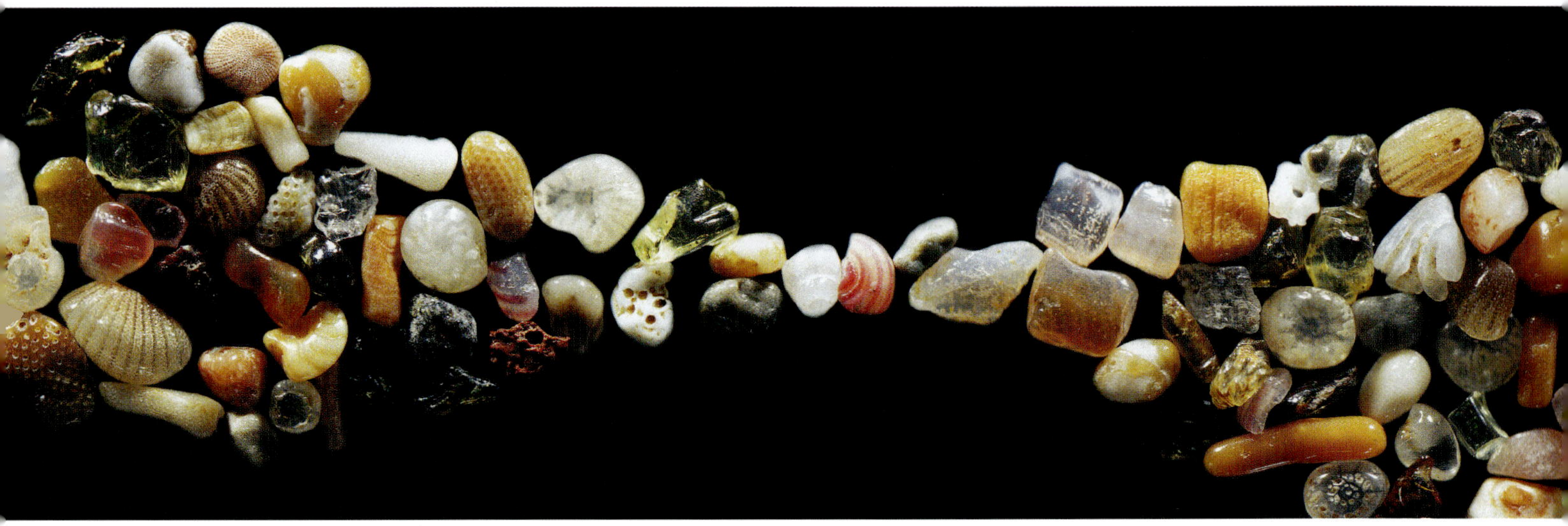

Sand grains from Ka'anapali, Maui, were herded into position on a microscope slide using acupuncture needles. The image is compiled from a series of eight individual images taken side by side (magnification 35x).

a grain that is outstanding, I carefully extract it from the pile using a very fine acupuncture needle.

It can take hours to fully examine a thimbleful of sand. It feels like I'm mining for gold and jewels but on a very small scale. This is one of the innocent joys of nature that I get to experience every time I look at a new sample of sand from somewhere in the world.

Once enough grains have been harvested, they are then laid out on a clean glass slide. I use an acupuncture needle to herd the little grains of sand into their final positions. This part of the procedure is the most difficult and requires a steady hand and a good deal of patience. Falling dust, a breath of wind, and static forces are the enemies. After the grains are in place and any dust particles have been cleared away, the cameras are attached to the microscope to record a series of 3-D photographs. It requires anywhere from six to twelve shots taken side by side to capture the whole scene. The individual photographs are then brought together in one large, seamless image using Photoshop.

In an alternative method, I take a single photograph of each individual grain of sand and then later arrange the sand grains electronically using Photoshop, rather than physically manipulating each grain of sand on the microscope slide. The individual sand grains are photographed on a black background in order to cut-and-paste without making them look as though they've been artificially extracted.

To make the background black, I put the sand on a glass slide. Highly oblique reflected light is produced by two large fiber-optic light sources built into the microscope. The lighting angle creates dramatic highlights and shadows. Since there is a dark empty space below the glass slide, it appears as though the sand is floating on a black background.

Most of the images in this book were created in the following way: First, I neatly spread out each sample of sand on a glass slide, and then I take a microscopic journey of one square inch of sand, grain by grain. When a particular group of sand grains catches my eye, I create a stack of through-focus images of those grains. The stack of images is then put together into a single, extended depth of focus image, where both the foreground and the background are in sharp focus. The majority of

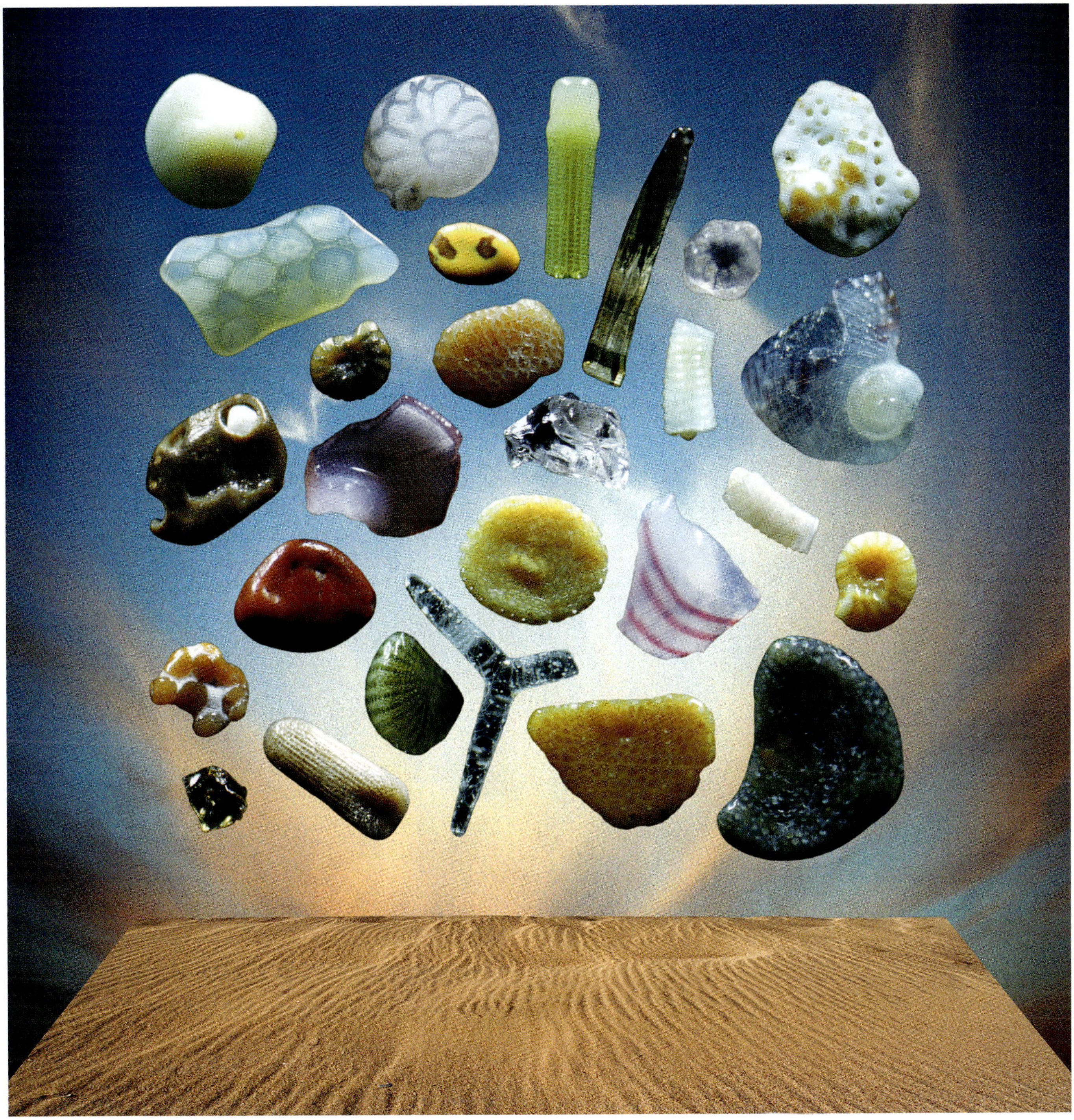

Maui sand grains appear to float above a beach in a montage made of twenty-seven individual photographs put together in Photoshop (magnification 90x).

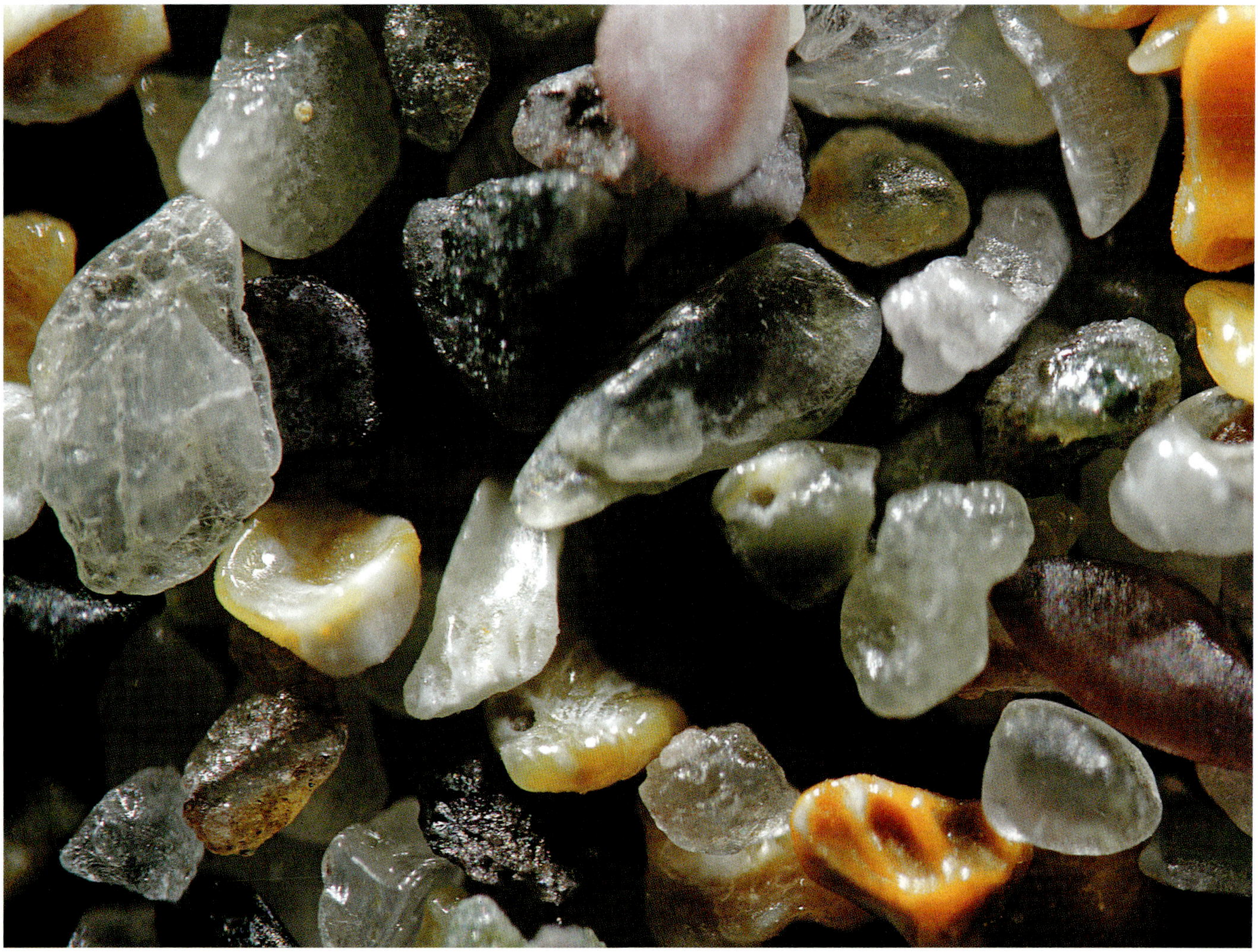

Sand grains from Lahinch, Ireland, are illuminated using two different techniques. *Above*: Very oblique reflected light creates one version of reality. *Opposite*: Transmitted light penetrating the grains creates a very different view (magnification 125x).

images in this book are from these random samples of sand, rather than a preselected collection of sand grains that had been hand-placed on a glass microscope slide, as previously described. Although I usually use fiber-optic light guides to illuminate the sand grains, I occasionally use transmitted illumination. Transmitted illumination is light that *passes through* the specimen that is then observed, as opposed to reflected illumination, which is light that is *bounced off* the specimen. I also employ crossed polarizing filters when using transmitted light. Polarized, transmitted light penetrates some of the sand grains, creating a different view than traditional reflected light.

Notice how different the same sand grains appear when illuminated in different ways! It's amazing what you can learn by shedding light on something from a different angle.

When I show my photographs to people, they often ask, "Is that what it really looks like?" That's an

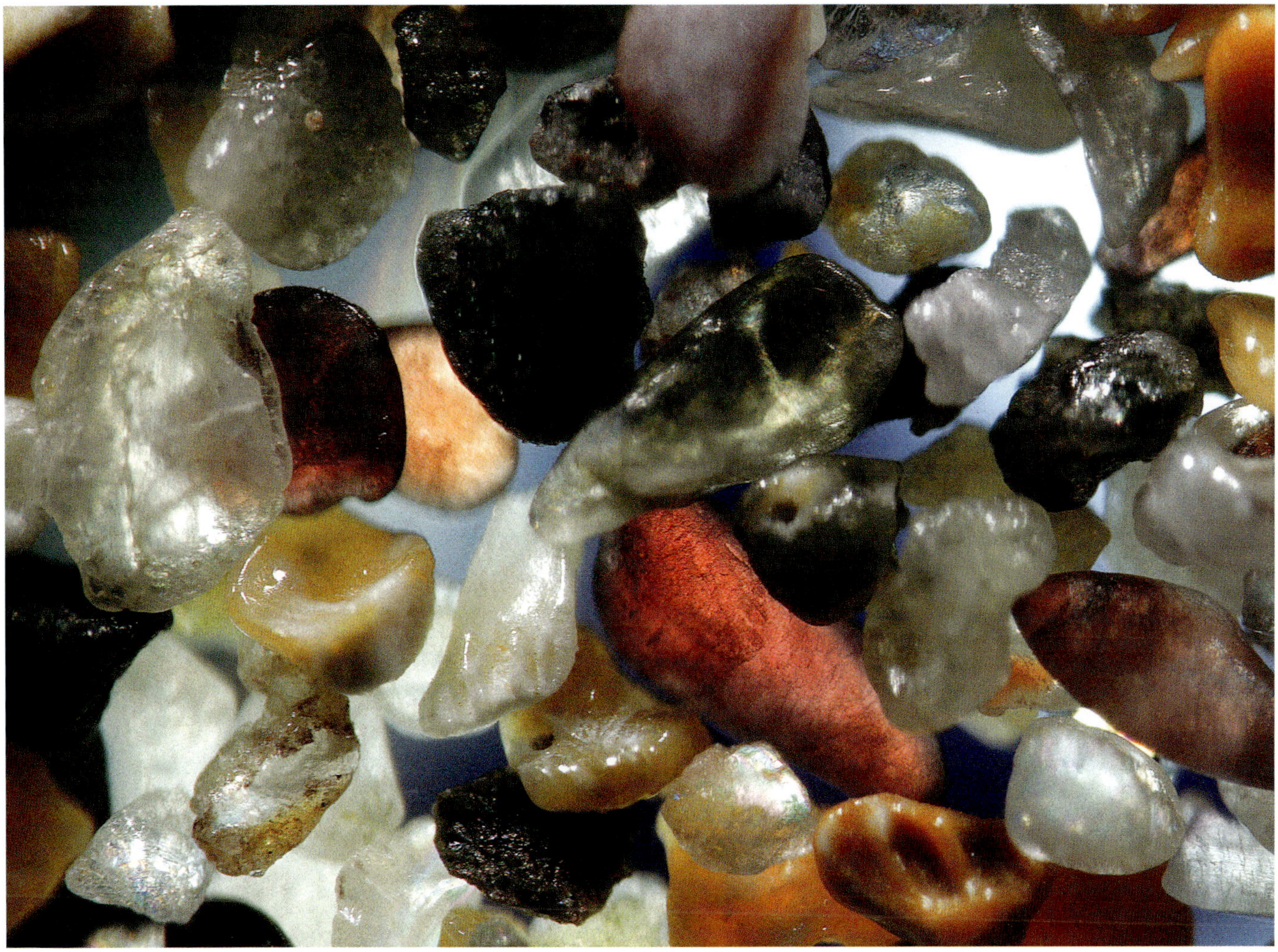

interesting question. There are dozens of different types of microscopes, and each one produces an image of the specimen in a different way, providing different information about the object.

So I guess there is no answer to the question "Is that what it really looks like?" Because the answer always depends on how you observe, measure, and categorize things.

Exploring the world through microscopes has taught me that we can't know everything about even a simple object. As a scientist, I'm trained to look at things from as many points of view as possible and then to build a model of the object in my mind in order to understand it better. That's an important lesson: There are multiple ways of looking at reality. Each tool of observation we have at our disposal produces a different result. With our nose, we smell an object; with our eyes, we see an object from a particular angle; with our skin, we touch the object; we may even listen to the object or taste it; and, in the end, we will label the object and create a model of it in our brain. Is that really what it is? No, certainly not. There are a thousand and one facets to every thing and every event—our consciousness cannot grasp them all. We are forever seeing the world from our own point of view. That's the inescapable reality of life.

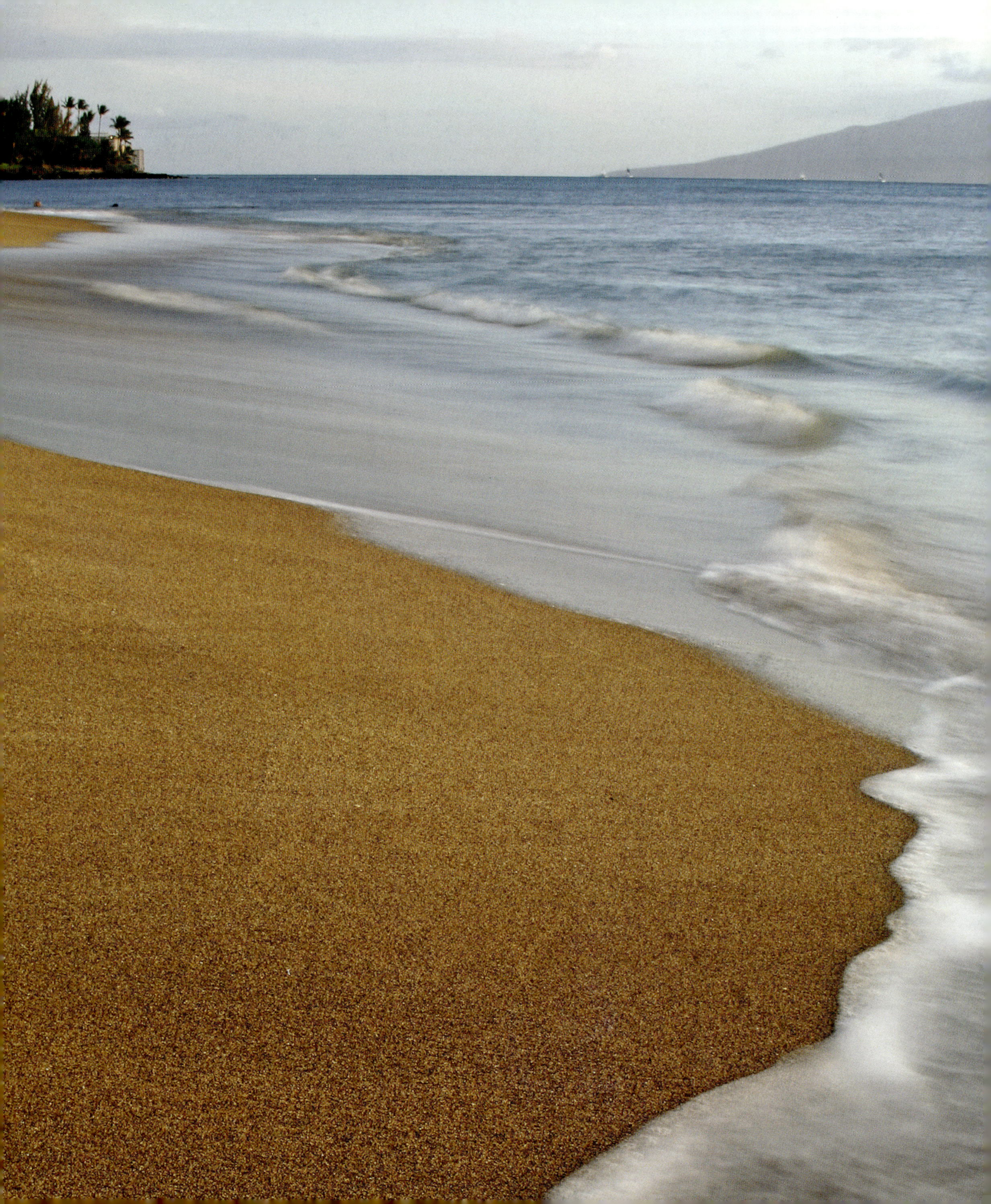

Chapter 3

The Nature of Sand

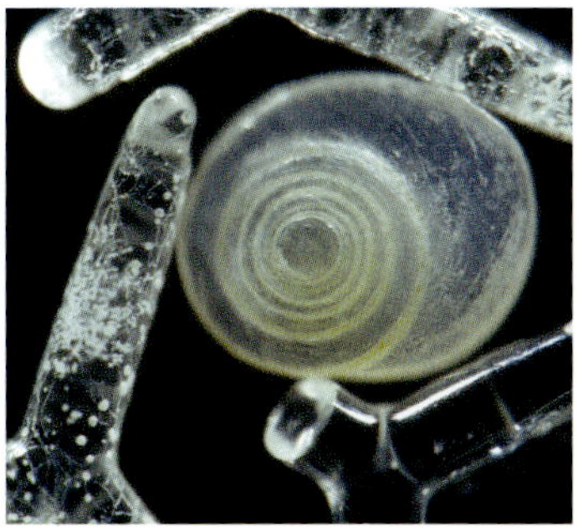
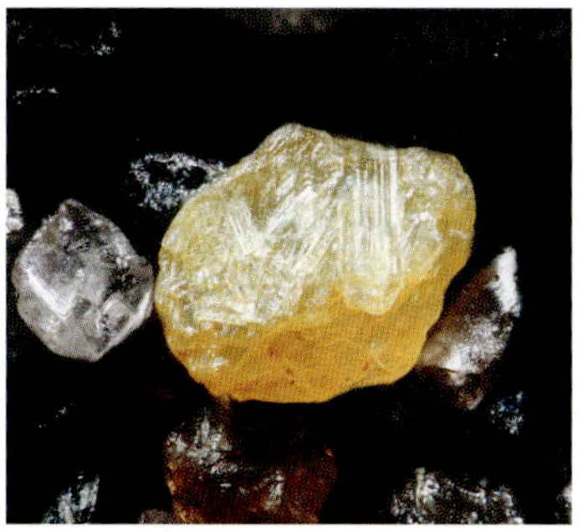
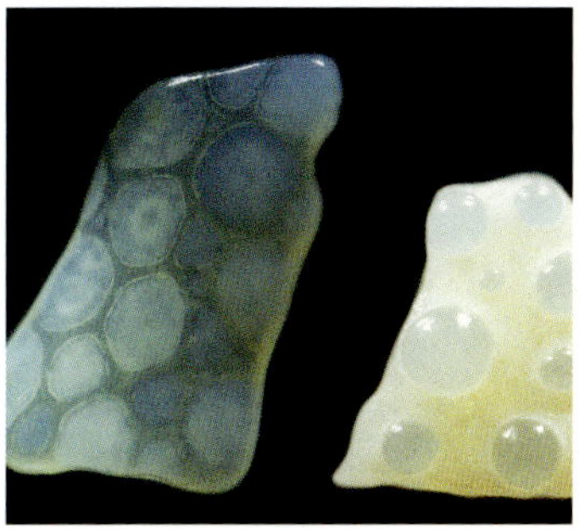

My fascination with sand fueled a desire to learn more about it. I have discovered several engaging features about sand grains. The most intriguing characteristics are, first, how diverse they are from each other and, second, how beautiful each individual grain is when examined through the microscope. One stunning example of the individuality of grains is the star-shaped sand grains on the beaches of Okinawa, Japan. In 2007, I received a container of this remarkable sand and I was amazed by what I saw. Each grain of sand looks like a puffy white star studded with little pearls. Actually, they are forams, which are single-celled, amoeboid protists that make microscopic shells.

I want to take you on a journey from the viewpoint of geology, mineralogy, and oceanography, so that we can more fully appreciate the humble and overlooked grain of sand. Sand is first distinguished by its size, which can vary from 2 millimeters in very coarse sand to 1/8 millimeter in very fine-grained sand. Most good sandy beaches have an average grain size of 1/2 to 1/4 millimeter. When the particles have eroded to smaller than 1/16 millimeter in diameter, the material becomes mud or silt; when it erodes to particle sizes of just a few thousandths of a millimeter, it's considered clay.

Waves lap on Kahana Beach, Maui, at sunset. *Stephanie Coffman, Shutterstock*

Singing Sand

Imagine you're in the desert surrounded by sand dunes with nobody in sight. You begin to hear eerie sounds coming from the earth. At first it sounds like a freight train, but there aren't any train tracks for a hundred miles. Then you think it sounds more like thunder in the distance, but there are no clouds in the sky. So then you think maybe it's an airplane, but no, that's not it either.

As it turns out, it's the sand dunes making the noise. Middle Eastern and Chinese societies have known about this peculiar phenomenon for over a thousand years. There are only a few dozen places on Earth that boast singing, booming, or barking sands. A little more than a half dozen of these sites are in the United States—places like Eureka Dunes, California; Barking Sands at Maria, Kauai, Hawaii; and Sand Mountain, Nevada.

In a related phenomenon, if the conditions are just right, sand that is walked upon can make squeaky noises. However, this occurrence is not as rare as singing or booming sands.

The conditions necessary to produce singing and booming sand are quite stringent. The sand is usually very dry, with less than 1 percent wetness. Also the sand must have a medium grain size and be well sorted for consistent size, so that all the grains have about the same diameter. The grains are quite spherical in shape with semipolished or frosted surfaces when viewed through a low-power microscope. However, unevenness in surface roughness on the electron microscope level is an important factor for acoustic quality. The pitch of the sound is related to grain size, while the volume of the sound is related to surface texture.

And finally, wind is the driving force in producing singing sand dunes. Quartz is almost always the material that makes up singing sands; the one exception is the singing sand in Hawaii, which is made of calcium carbonate.

Most singing and booming sands produces a very low-frequency sound, limited to about 50 to 80 hertz (cycles per second). That's an extremely low-pitched sound for the human ear to perceive. There are also harmonic peaks at 100 and 180 cycles per second. People report that it sounds like an airplane. The very low-frequency vibrations of booming sands can make the sand tremble and ripple, and one can literally feel the vibration.

Dunes can also make musical sounds when an avalanche occurs. This shearing of sand produces a sound more like whistling or booming. The sounds only last a second or so, and they are higher pitched, in the range of 500 to 2,500 cycles per second.

Here is a firsthand account from a person who experienced singing sands in western Egypt:

> *On two occasions it happened on a still night, suddenly there was a vibrant booming so loud that I had to shout to be heard by my companion. Soon other sources, set going by the disturbance, joined their music to the first, with so close a note that a slow beat was clearly recognized. This weird chorus went on for more than five minutes continuously before silence returned and the ground ceased to tremble.*

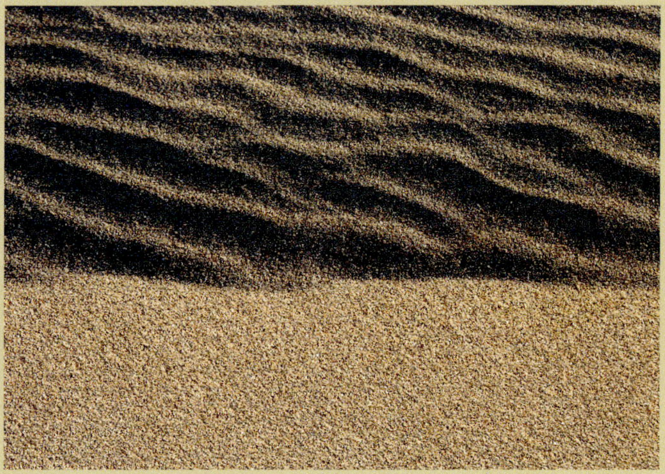

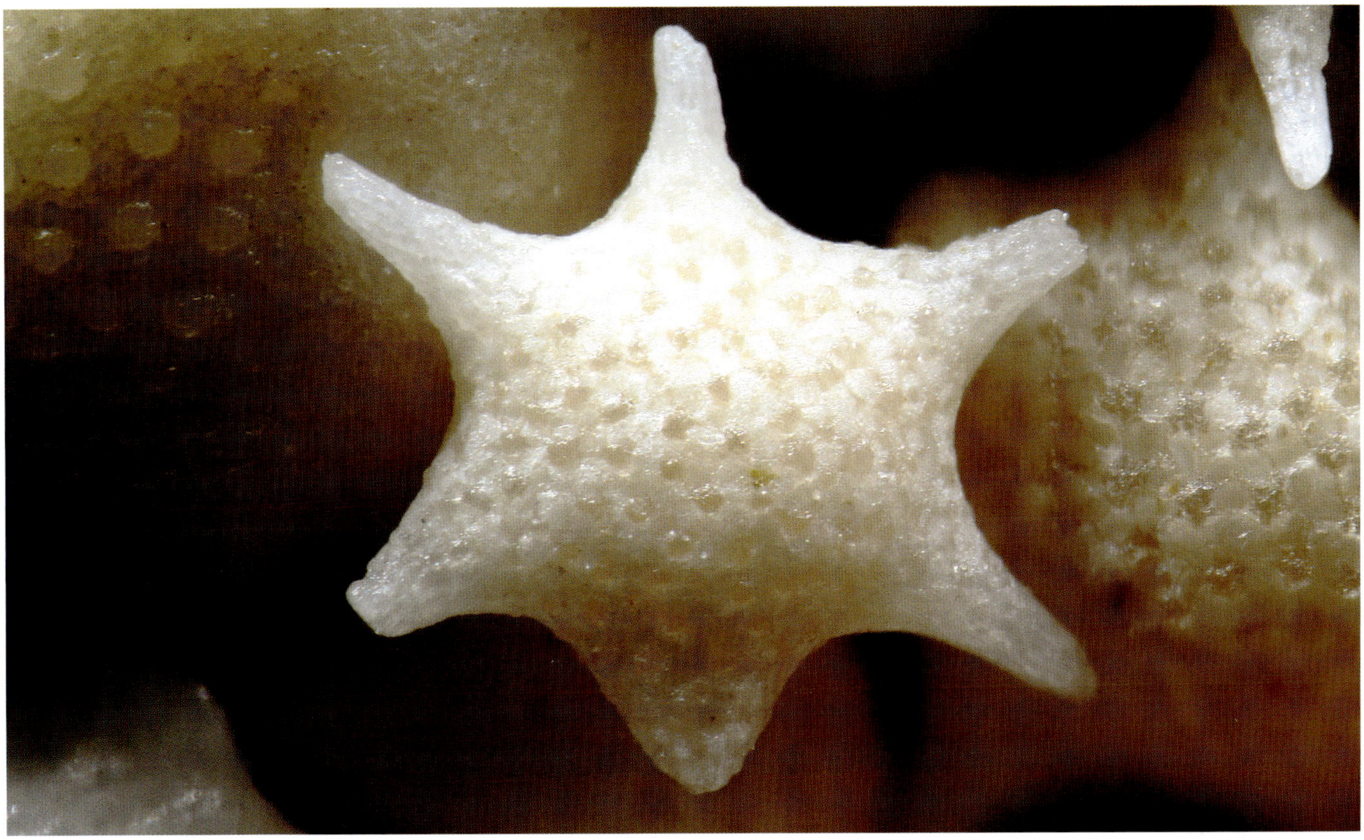

A star-shaped grain of sand from a beach on Taketomi Island in Okinawa, Japan. The sand contains the remains of microscopic shells known as forams (magnification 75x).

How many grains of sand are there in the world? The famous astronomer Carl Sagan said that the total number of stars in the universe is greater than all the grains of sand on all the beaches of the planet Earth. These numbers are difficult to accurately estimate, and several approaches have been used to find the answer. All estimates are based on assumptions that are not well established, such as the average size of each grain of sand, the breadth and thickness of the sand layer that covers the beaches, and even the exact length of shorelines in the world. By one estimate, there are 4,800,000,000,000,000,000,000 grains of sand on the world's beaches; that's 4,800 billion billion grains of sand, or 4.8×10^{21}. This estimate is based on the following specifications: an average grain size of 1/4 millimeter and an average of 50 meters of sandy beach, 1 meter deep, along 1.5 million kilometers of shoreline, with the sand grains perfectly packed together.

How many stars are there in the universe? Only about 6,000 stars are observable with the naked eye on a dark evening, all of which are nearby stars in our own galaxy, the Milky Way. However, survey data from the Hubble Space Telescope suggests that the universe contains about 130 billion galaxies. And the average galaxy, such as our own Milky Way, contains about 400,000,000,000 stars. That works out to about 52,000,000,000,000,000,000,000 stars in the universe; that's 52,000 billion billion stars, or 5.2×10^{22}. So Professor Sagan might be correct that the stars in the heavens outnumber the grains of sand on all the world's beaches.

Apart from sand on beaches, a massive amount of sand covers Earth's deserts and the margins of its oceans, lakes, and rivers.

The Creation of Sand by Erosion

The planet is continuously changing. Mountains rise, then erode and weather away. Mineral sands form from the mechanical and chemical breakdown of rocks. As solid and permanent as the granite mountains in Yosemite National Park appear, they are continuously weathering away. The massive granitic rocks have been exposed to the light of day only because the surrounding material has completely eroded, leaving the denser, more resistant rock behind—for the time being.

Consider the Appalachian Mountains versus the Rocky Mountains. The Rockies are relative newcomers, formed between 140 million and 65 million years ago. They are tall and jagged, while the older Appalachians are heavily eroded. Hundreds of millions of years ago, that region of the world was connected to the Baltic Sea. At one time, it was near the center of the supercontinent Pangaea. Starting about 450 million years ago, a series of mountain-building events occurred, resulting in the creation and subsequent erosion of the Appalachians. Now they appear smooth and worn when compared to the younger Rockies. In fact, much of the beach sand on the East Coast came from the erosion of the Appalachians.

Beach formation begins with wind, rain, snow, ice, and glaciers wearing down mountains and eroding tiny bits of rock, which are carried to the sea by streams and rivers. Rivers are the medium via which 80 to 90 percent of all sand reaches a beach. Satellite views of river deltas reveal the enormous amount of sand that is continuously pouring into the ocean. The sand is carried away by waves and currents and takes up temporary residence along the local beaches. Floods can also move vast amounts of sand from the mountains to the sea.

Since most continental landmasses are made of granite, most sand is the product of eroded granite. When granite breaks down, it turns into quartz, feldspar, hornblende, mica, and minerals such as garnet, magnetite, and zircon.

In some cases, even man-made objects erode to become grains of sand. One such example has occurred on the Greek island of Delos. During the reign of the

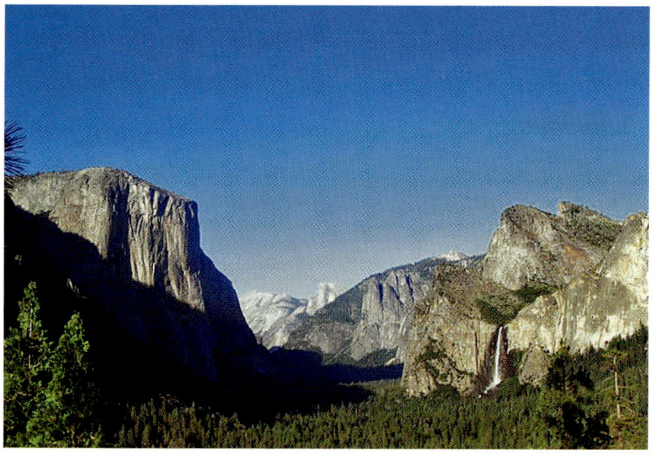

Most quartz sand is the end product of eroded granite rock, such as seen in the landscape of Yosemite National Park, California. *Wikimedia Commons*

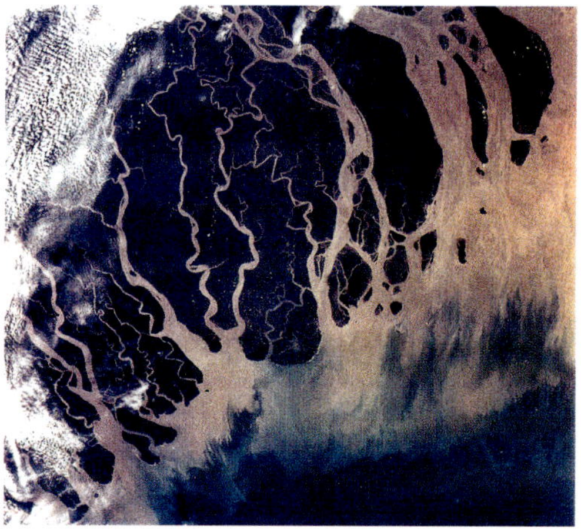

An aerial image of the Ganges River Delta shows the vast amount of sand the river continually washes to the sea. *Wikimedia Commons*

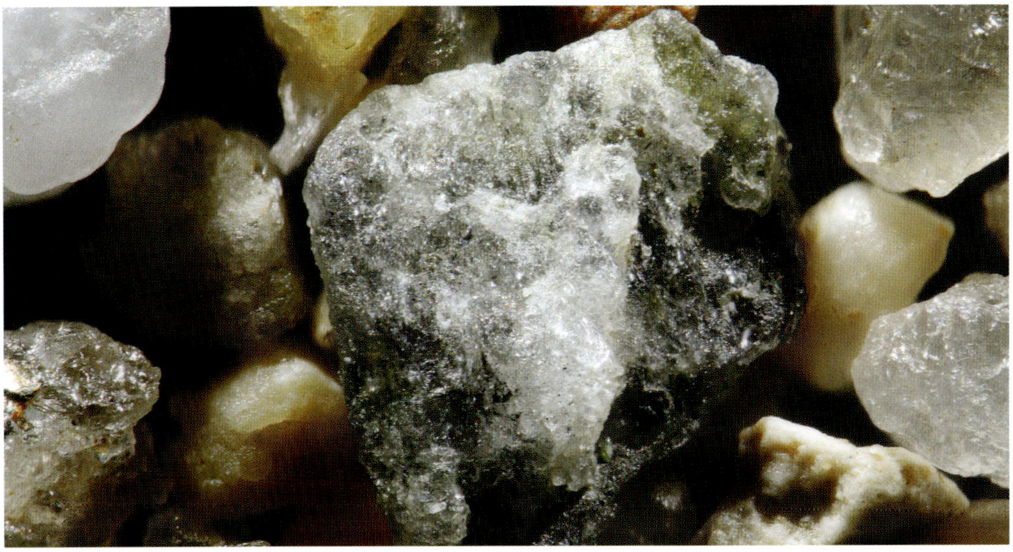

The sand on the beaches of Delos Island is peppered with different types of marble eroded from temples built by the ancient Greeks (magnification 130x).

ancient Greek Empire, Delos was the site of a vast number of marble temples. Delos had no indigenous marble, and a colossal amount of marble had to be quarried and brought to this small island to erect the temples. Over the years, the marble blocks have eroded and the local beaches have filled with tiny grains of sand made of beautiful nonnative bits of temple marble.

A much more alarming example of "sand grains" that originate from man-made objects can be found on trash beaches around the world. Unfortunately, we are still using our oceans as garbage cans. Out of sight, out of mind. But the wind, waves, and currents conspire to dump massive amounts of floating debris on beaches around the world. There is just such a trash beach at the southern tip of the Big Island of Hawaii, where litter of every description is piled high on what would otherwise be a beautiful beach filled with driftwood from the Pacific Northwest. Instead, this trash beach is heaped with rubbish, in particular, bits of plastic.

Every day, enormous amounts of tiny plastic pellets, used in the process of molding plastic products, escape from trains and warehouses. The plastic finds its way to the ocean via storm drains. This appalling environmental oversight severely impacts birds, which mistake the bits of plastic for food. Countless birds are killed yearly from eating plastic. Autopsies have revealed the stomachs of these birds to be filled with brightly colored pieces of plastic and other man-made debris. A 2006 study on Midway Atoll in the Pacific Ocean reported that plastic and other debris kill as many as 40 percent of albatross chicks, "their bellies full of trash."

We need to recognize that our planet is a wonderful living ecosystem and that the ocean is the cradle of all life on Earth. We should respect our oceans and our beaches. They are valuable resources that need to be protected from the forces of thoughtlessness, ignorance, greed, and complacency.

The Contents of Beach Sand

As we have seen, sand isn't made up of just little bits of broken rock. There are also lots of biologically derived components in sand. These biogenic sand grains are more numerous on island beaches or beaches adjacent to coral reefs or tide pools. When marine organisms die, they leave behind remnants of their calcified tissues, such as their shells, bones, and teeth. Biogenic materials represent the major ingredient of sand on some tropical beaches. Another source of sand is particles that have precipitated from minerals in solution, such as calcium carbonate.

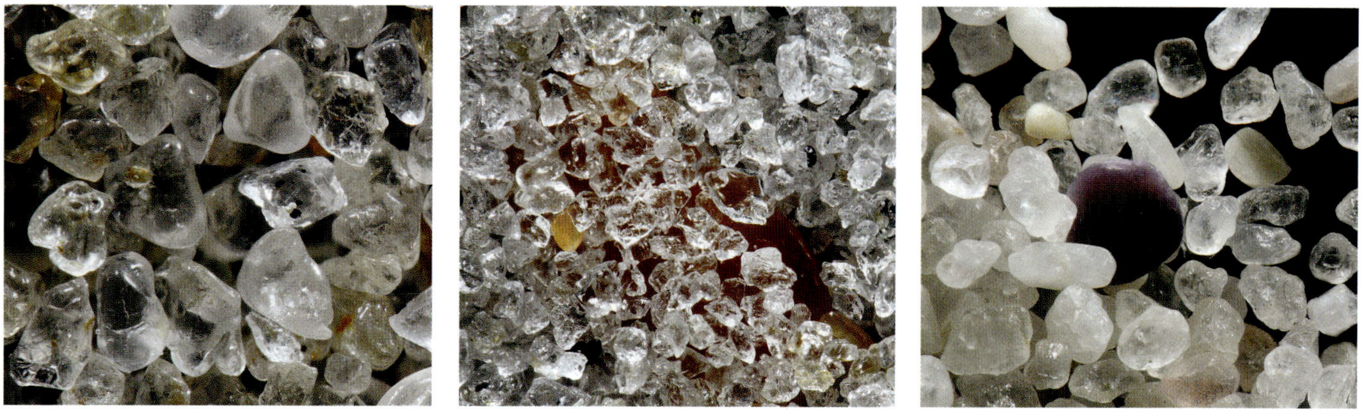

Sand on most continental beaches is made up primarily of quartz, which is relatively resistant to erosion. *Left*: A sample of quartz sand from Copacabana Beach in Brazil is smooth and rounded. *Middle*: Quartz sand from Florida appears jagged, indicating that the grains broke away from the parent rock relatively recently. *Right*: Quartz sand from Cape Town, South Africa, is worn smooth and likely broke off the parent rock several hundred million years ago (magnification 30x).

So sand can originate from three different sources: 1) rocks, which form mineral sands, 2) organisms, which form biogenic sands, and 3) mineral solutions, which form precipitated sands.

Beaches vary in color and texture according to the contents of their sand. The colors range from pure white to pure black, with a myriad of variations in between, including pure red and pure green. A microscope reveals the various components of sand. First, we'll look at the mineral components of sand formed from rocks. The chief mineral constituent of most sand is silica, usually in the form of quartz. Beautiful white-sand beaches are typically made of quartz sand grains.

Mineral Components

All mineral sands come from the erosion of rock. The mineral components of a sand grain provide clues to the origin of the eroded rock. These bits and pieces tell the story of the geological history of a particular beach.

There are three basic types of rock: igneous, sedimentary, and metamorphic. *Igneous* means "fire" in Latin, and igneous rocks are produced by the solidification of hot, molten magma. Magma that erupts from volcanoes cools quickly to form extrusive igneous rock. In contrast, magma trapped at great depths beneath the Earth's surface cools slowly to form huge granite masses known as intrusive igneous rock.

The second rock type, sedimentary, forms from the eroded rock material that is continuously deposited into streams, lakes, and seabeds. Layer after layer, the sediment builds up over time. Thousands or even millions of years of compression cause the sediment to naturally cement together and form sedimentary rock. Sandstone, often used as a building material, forms from sediment that contains quartz. Limestone forms from sediment that contains a high proportion of biological matter. It is often made of calcium carbonate that originates from evaporated lakes and seabeds that contained biological materials such as shells, fossils, and coral fragments. Limestone is used as a constituent of concrete.

The third rock type, metamorphic rock, started out as either igneous rock or sedimentary rock but has subsequently undergone a transformation due to extreme pressure and heat. Metamorphic rock is often banded, foliated, or irregularly colored, evidence of its abrupt change in structure or composition. Common metamorphic rocks include schist, composed of laminated, flaky parallel layers; gneiss, which is coarser than schist and has distinct banding derived from granite; and marble, derived from limestone. Metamorphic rocks can erode into sand, too; beaches such as those at the Isle of Shoal, Maine, and Plum Island, Massachusetts, are comprised of metamorphic minerals such as garnet and kyanite.

The sand on a tropical island differs significantly from the sand on a continental beach. This is because the rock underlying the ocean floor is mainly basalt, while the rock underlying the continents is mainly granite. So the mineral components of island beach sand are made up of basalt from the ocean floor, while continental beach sand is made up typically of quartz and feldspar derived from the eroded granite of the local bedrock. There is no quartz in the sand on Hawaii because there is no granitic rock there.

The Earth's surface is of made up of enormous plates of igneous rock called tectonic plates. About ten major plates and a couple dozen minor plates cover the surface of the Earth. These huge landmasses float on top of the Earth's mantle, which is made of molten rock. Below that is the Earth's core, which is under great pressure, and its temperature is more than 5,000 degrees Celsius. The convection of heat from the core keeps the molten magma moving.

Tectonic plates are essentially sheets of cooled rock. They shape the Earth's continents and seafloors. On a geological time scale, these plates are continually moving. They move one to ten centimeters per year, which is about the rate that our fingernails grow. The rate of spreading along the Mid-Atlantic Ridge averages 2.5 centimeters per year. That's 2,500 kilometers in 100 million years. During the Earth's 4.5-billion-year history, continents have been created and destroyed many times over. About 250 million years ago, all the continents we know today were part of a single massive supercontinent, called Pangaea. Then, 200 million years ago, Pangaea began to split and drift apart, forming the continents we know today. Portions of the ancient split between the continents can still be seen in the way Africa and South America fit together like jigsaw puzzle pieces.

The boundaries between plates create active geological areas. At some boundaries, new crust is being formed, such as in the middle of the Atlantic Ocean. There, a ridge of underwater volcanoes spews out molten magma, which quickly cools and hardens into rock. The creation of new crust pushes the two adjacent plates apart, which results in the spreading of the seafloor. At other plate boundaries, such as in California and Japan, the tectonic plates sliding against each other cause massive earthquakes.

Where tectonic plates push together, mountain ranges can form, such as where the Indian-Australian plate has collided with the Eurasian plate, resulting in the formation of the Himalayas, home to the world's tallest acknowledged mountain, Mount Everest. (If you measure from the seabed rather than sea level, Mana Loa on Hawaii is the tallest mountain.)

At yet another type of boundary, where an ocean plate collides with a continental plate, the denser ocean plate slides under the continental plate. This is known as a subduction zone. The edge of the ocean plate is pushed downward and part of the Earth's crust is recycled back into magma. The Andes Mountains in South America and the Cascade Range in the Pacific Northwest are both regions where ocean crust is being subducted under continental plates. The 1980 eruption of Mount St. Helens is an example of the turmoil these geological events can unleash.

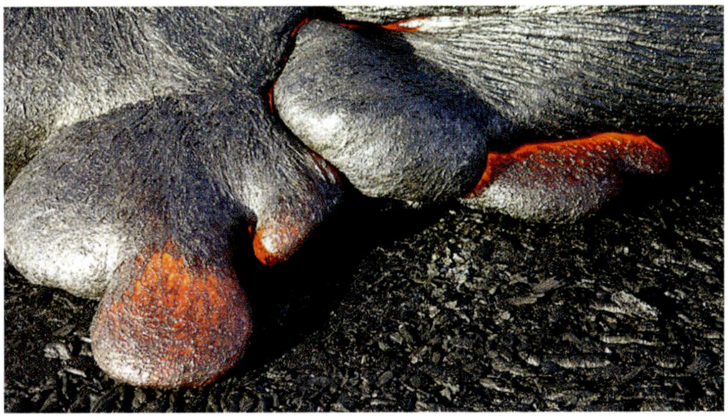

Lava in Hawaii cools to form basalt rock, which erodes into grains of sand if exposed to the continual pounding of the surf. *Wikimedia Commons*

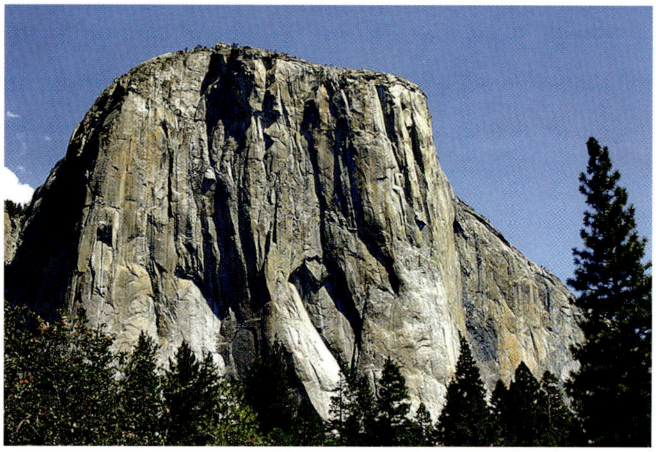

Granite in Yosemite National Park slowly solidified from hot magma beneath the Earth's surface. Uplift, along with the erosion of surrounding materials, exposed the granite rock. *Wikimedia Commons*

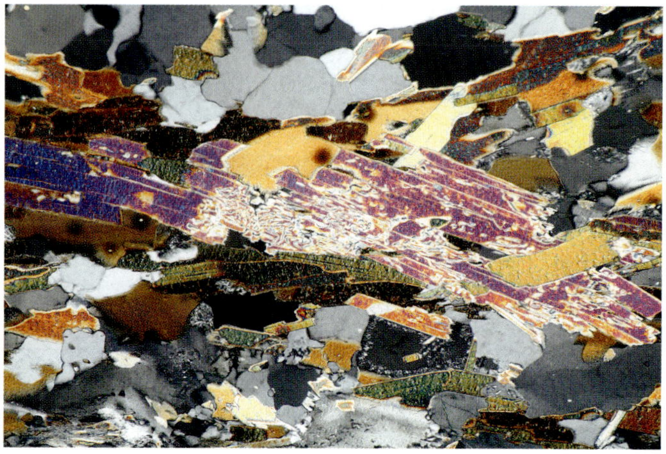

A microscopic image of a slice of rock shows that individual grains are separated by clear boundaries. As rock erodes, grains work free to become sand grains while others break down to silt- or clay-sized particles (magnification 85x).

To review, there are two types of tectonic plates: The continental plates comprise mainly granitic rock, which is rich in silica, whereas the oceanic plates comprise mostly basaltic rock, which is denser. Basalt also contains silica, but it is richer in magnesium, iron, and other heavier elements.

Granite is a light-colored, coarse-grained rock formed deep within the Earth, exposed only after massive upthrust and erosion. It cools slowly because the Earth's insulation keeps the heat from being conducted away quickly. Granite is rich in feldspar and quartz. One of the most famous granitic rocks is Half Dome in Yosemite National Park, California. In contrast, basalt is a dark, fine-grained rock that cools relatively quickly because of its direct exposure to water or air.

Does a grain of sand exist before it breaks off the parent rock? Yes. If you examine a thin slice of rock under the microscope using crossed polarizing filters, you can clearly make out the grain boundaries between the various crystals and minerals. As the rock weathers, the grains work loose and wash away as sand.

The shape of each grain of sand holds a clue to its history. For example, the degree of roundness is an index of its age. All sand grains that are formed by weathering of rock start out angular and become more rounded as they are polished by abrasion from wind, water, ice, and tumbling. Quartz grains are the most chemically stable and resistant to wear; they can survive many cycles of erosion, burial, compression into sandstone, uplift, and re-erosion into sand again. Recycling time takes about 200 million years, so a quartz grain first eroded from granite a billion years ago may have gone through five cycles of burial and re-erosion to attain its present condition. A newly formed quartz grain has sharp edges, whereas a recycled quartz grain is rounded. Feldspar grains can also survive recycling but not as hardily, so sand that has been recycled a few times consists mostly of quartz.

The erosion of sandstone creates a staggering view in places such as Sedona, Arizona, the gateway to the Grand Canyon. Hundreds of millions of years ago, long before the forces of nature gouged away the Grand Canyon, this entire region was situated at the northwestern edge of the United States. It was actually underwater and contained a huge sandy seabed. The once-sandy bottom compressed into sedimentary rock and the land rose up from tectonic forces. Beginning 160 million years ago, these tectonic forces lifted up the

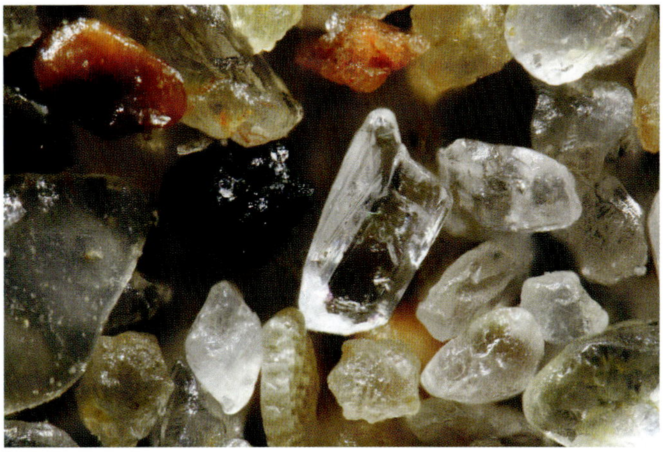
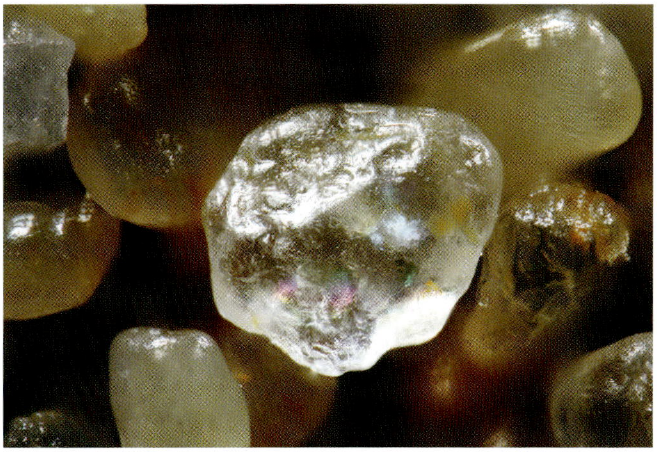

The crystal structure of (*left*) a quartz sand grain from Leisure Bay, South Africa, is still recognizable, whereas that of (*right*) a quartz grain from Maitland, South Africa, is eroded beyond recognition due to several cycles of cementation and erosion (magnifications 85x).

Colorado Plateau; the Rocky Mountains formed over the next 120 million years. Lots of rain and snow fell on the young Rockies, eventually giving rise to the Colorado River. Originally, the Colorado did not flow into the sea. About 5 million years ago, when the Gulf of California was opened up by the action of tectonic forces at the San Andreas Fault, the Colorado River began to flow into the Sea of Cortez. Torrents of water from the Colorado River, in combination with uplift from tectonic forces, gouged the Grand Canyon. Today, the walls of the Grand Canyon expose two billion years of Earth's evolution, which stands as a testament to the geological and biological history of our planet.

The magnificent spires of red rock around Sedona remain as evidence of these colossal erosion events, standing like majestic cathedrals in nature. The spires are beautifully decorated with the history of the Earth written in stone. The awe-inspiring setting is a dramatic reminder of the recycling process that all of nature experiences, from a tiny grain of sand to the entire Earth itself. Sand grains from this region of the globe illustrate the sheer beauty of the tiny bits and pieces that form our world.

Some sand contains a large amount of silica, a mineral that occurs as quartz. If the sand contains a high percentage of silica, it can be used to make glass.

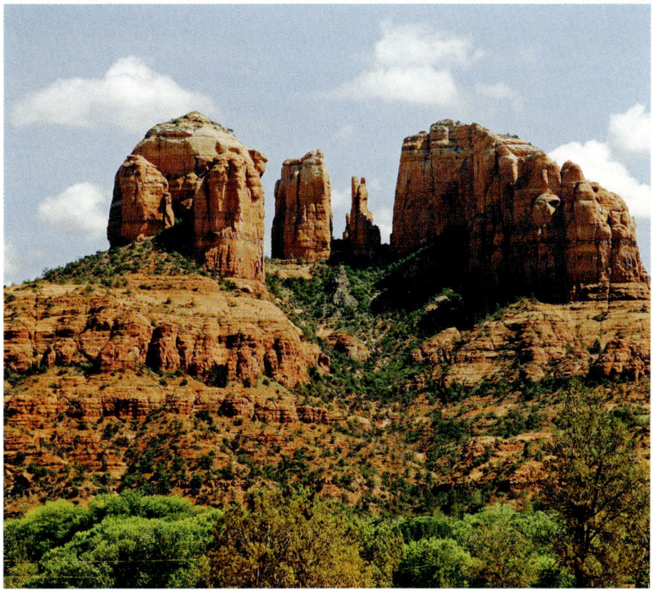

Hundreds of millions of years ago, regions like Sedona, Arizona, and the neighboring Grand Canyon were an underwater seabed. Layer after layer, the sandy bottom was compressed by overlying sediment. Then the terrain rose up from the seabed by tectonic forces, and the strata hardened into sandstone. The red color is due mainly to the high iron content, which oxidizes with air to form iron oxide. Cathedral Rock stands as one of Mother Nature's great sand castles.

In our day and age, we take glass for granted, but the invention of it truly changed the quality of human life. Pure quartz sand is mined to make optical glass. When an electric current is passed through quartz crystal, it vibrates at an exact frequency that is so reliable that it's used to control clocks and watches, producing a very high level of accuracy. Also, high-quality quartz sand is used in the silicon wavers that are utilized to fabricate microchips and other electronic devices. Silicon has also found a use in the manufacturing of micro-electro-mechanical systems. This form of nanotechnology is currently used in modern cars for the sensors that trigger airbags and for air pressure gauges built into some tires.

Magnetite is the most common magnetic mineral on Earth. Lodestone, which is made of magnetite, was used in early magnetic compasses for ship navigation. A form of iron oxide, magnetite is a frequent constituent of sand. Black streaks in beach sand are sometimes mistaken for oil streaks, but they are actually accumulations of magnetite. The tiny little magnetic sand grains naturally attract one another. I often notice magnetite in particular because it sticks magnetically to the acupuncture needles I use to sort out individual grains under the microscope.

Feldspar, mica, hornblende, and various minerals such as garnet and zircon are other constituents of eroded rock that become grains of sand. These minerals are gemlike when viewed through the microscope.

Basalt is a dark-colored igneous rock with less than 52 percent silica. Basaltic lava has a low viscosity, so it flows easily. Basalt is composed of iron-rich minerals, including dark-colored pyroxene and amphibole minerals. Olivine, named for its greenish-yellow color, may or may not be present in basaltic flows; however, olivine is a significant mineral in the basaltic flows in Hawaii and is the dominant mineral in the green-sand beach at South Point on the Big Island.

Biogenic Components

The sand on some beaches is packed with the remnants of biological organisms. These biogenic components of sand are undoubtedly the most

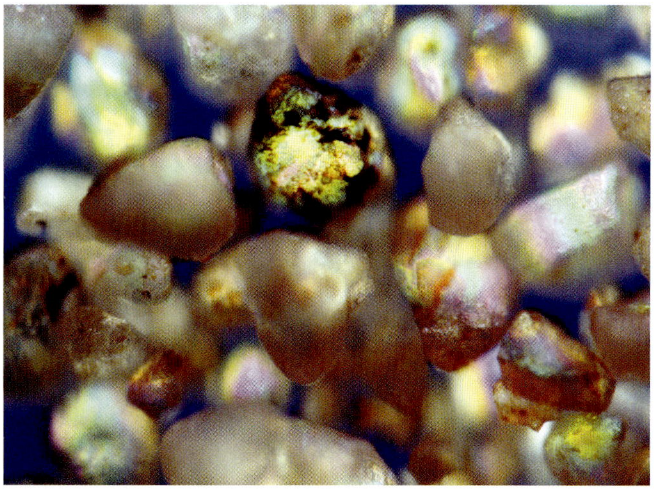

Sand grains from Moab, Utah, are seen with crossed polarizing filters. They have eroded into rounded shapes from their original jagged forms. The grains have likely been recycled many times from sand to sandstone and back to sand again (magnification 60x).

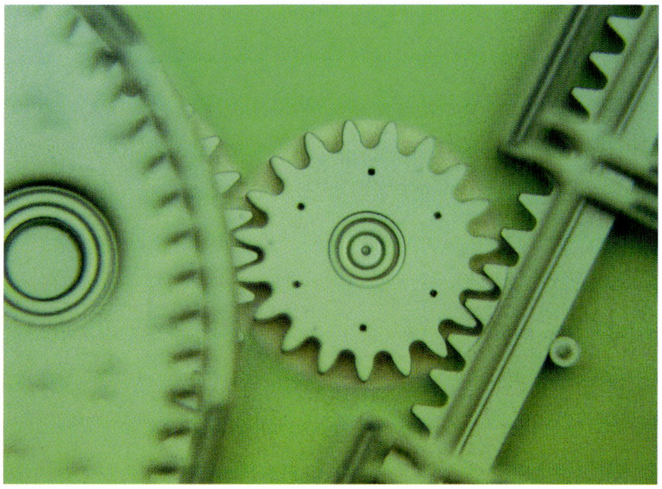

A micro-electro-mechanical system is smaller than a grain of sand. It's etched from silicon, which is made from quartz sand (magnification 70x).

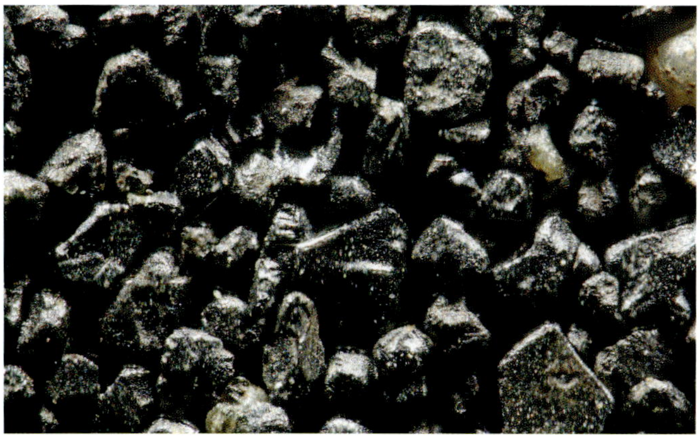

Magnetite, also know as lodestone, is a naturally occurring magnetic material (magnification 35x).

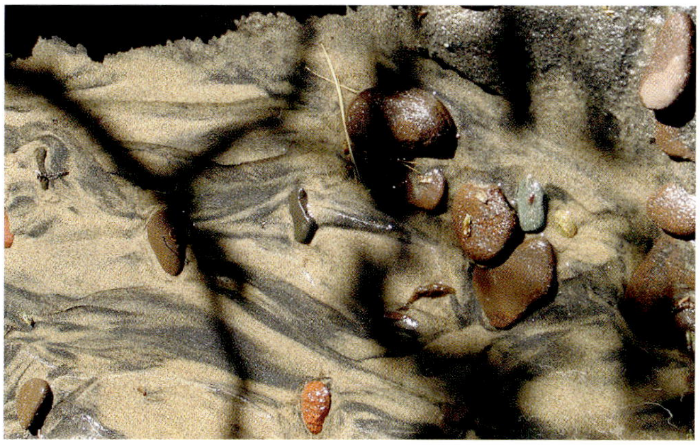

Magnetite in sand can sometimes look like an oil streak. The magnetite accumulates in patches because of its density. *Kate Hintz*

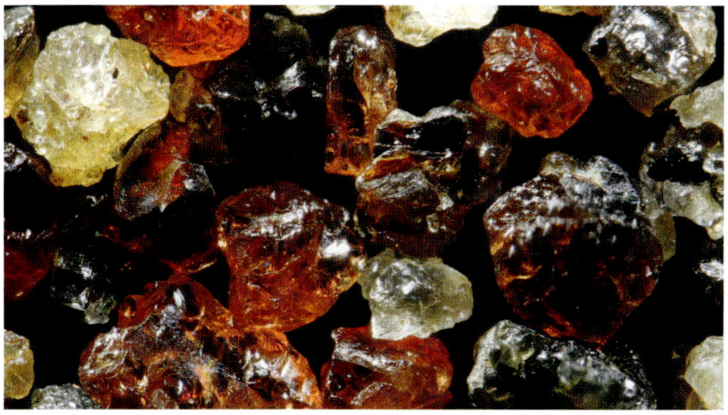

Sand grains made of dark minerals, garnets, and quartz were collected from High Bluff Island, Northwest Territories, Canada (magnification 85x).

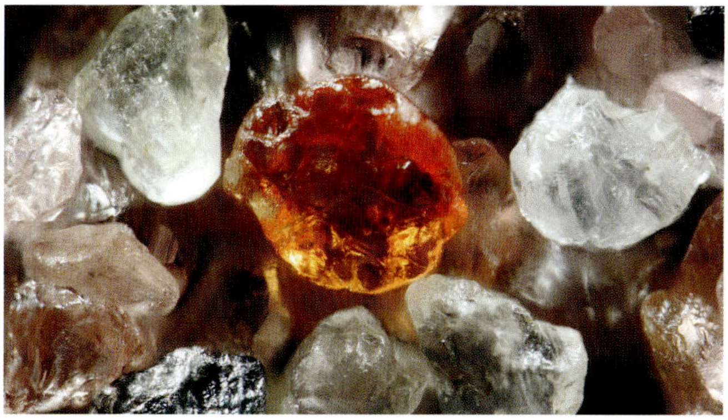

Sand grains from Plum Island, Massachusetts, are made of garnet, magnetite, quartz, and green olivine (magnification 85x).

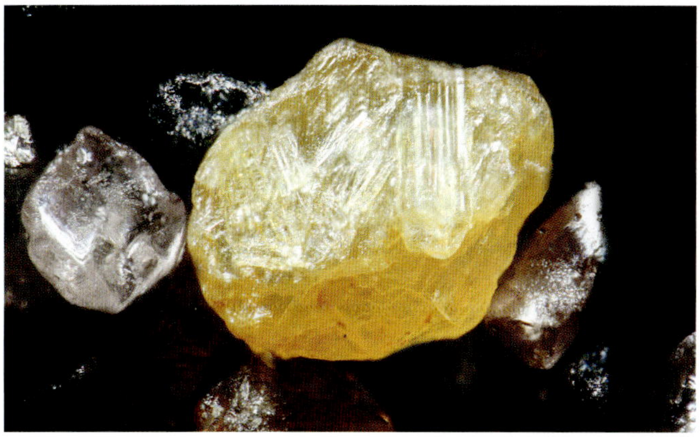

A tiny yellow crystal is found as a grain of sand from Plum Island, Massachusetts (magnification 110x).

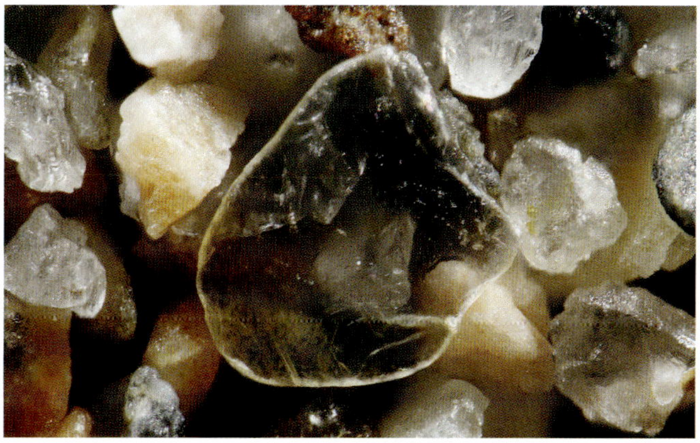

A translucent grain of mica from the South China Sea (magnification 110x).

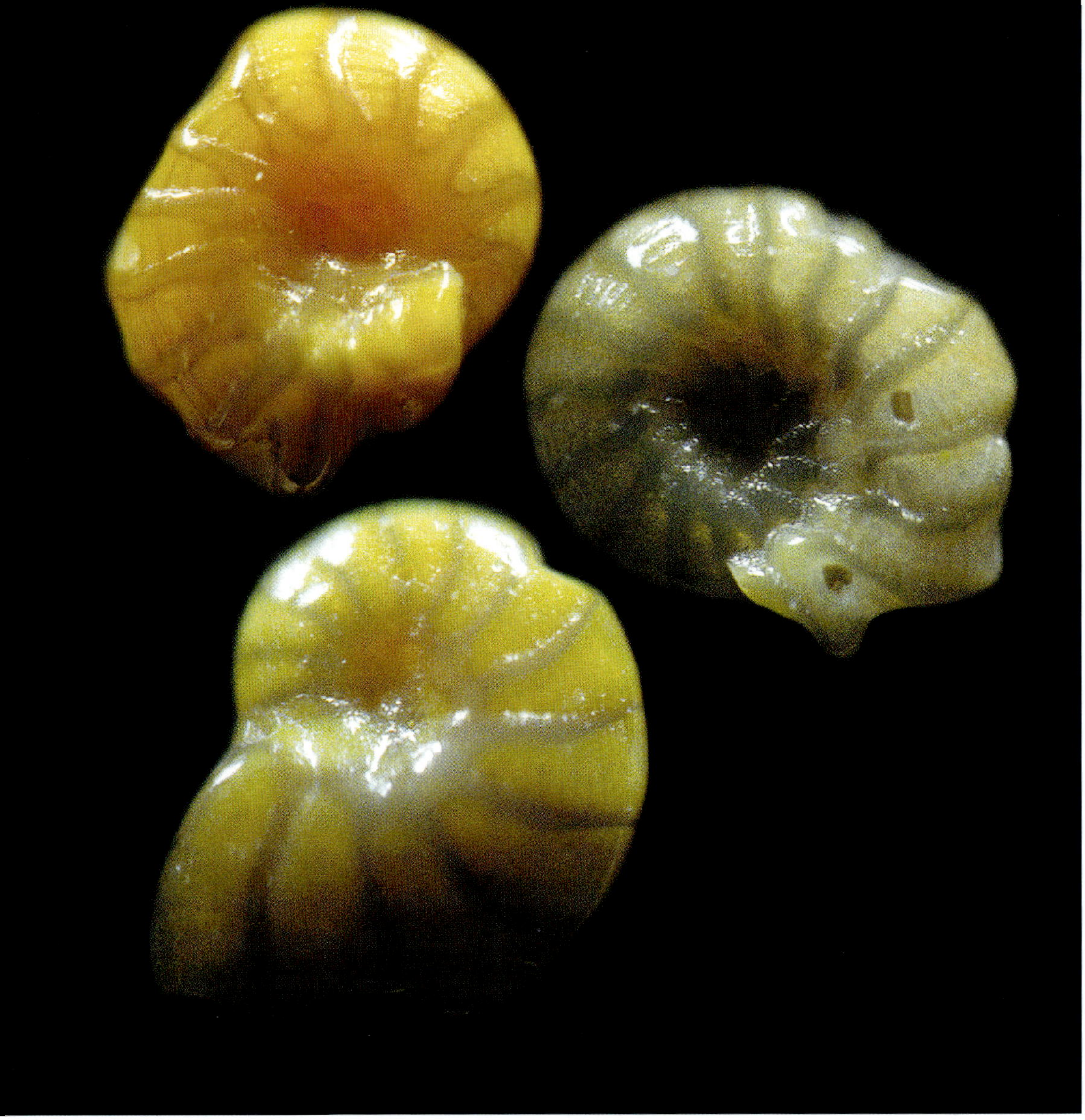

Microscopic shells from Maui are a striking example of the beautifully preserved microorganisms found in tropical beach sand. Only two small holes have pierced the integrity of the shells (magnification 130x).

Below: A perfectly formed microscopic mollusk shell from Cable Beach, Australia, has become a small grain of sand (magnification 85x).

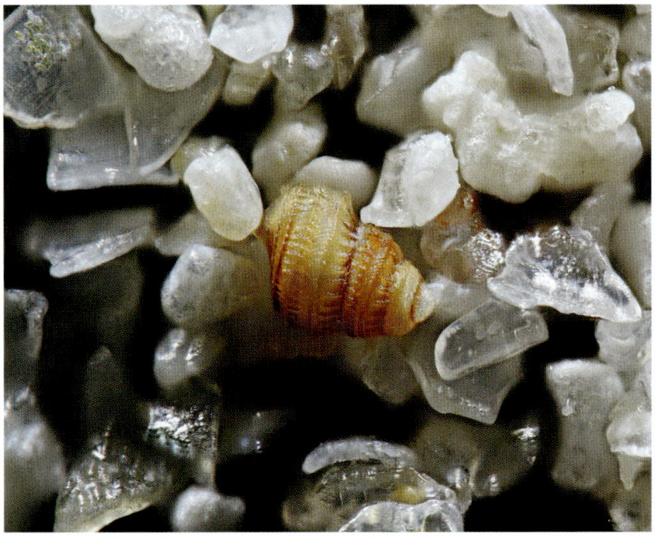

interesting and varied of all sand grains. Biogenic sand forms from the broken bits and pieces of the skeletons of marine life, such as shells, corals, sea urchins, sponges, fish bones, shellfish, starfish, sand dollars, and snails. In addition, beach sand can contain the magnificent skeletons of microscopic organisms. It is amazing how these delicate, minuscule forms of life survive the ravages and torrents of the surf.

Some of these shelled creatures are so small that they're invisible to the naked eye. The oceans are filled with tiny, single-celled, amoeboid protists called foraminifera, or forams. Today, an estimated four thousand kinds of forams live in the world's

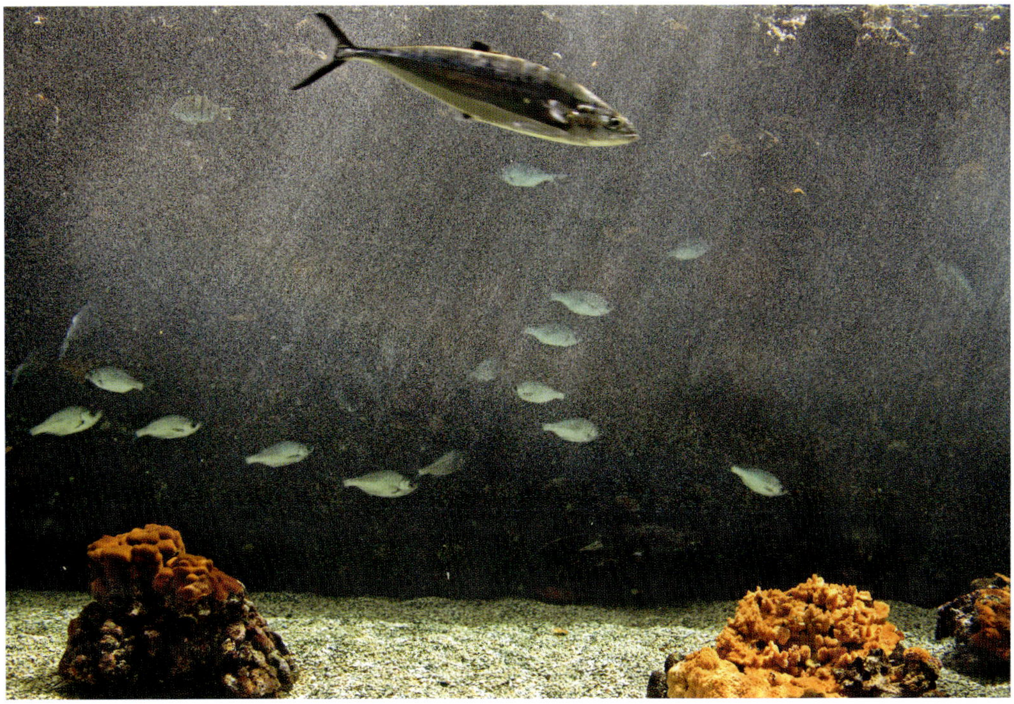

The ocean is thick with floating microscopic organisms called forams, easily seen when illuminated by shafts of sunlight. Forams combine to make up plankton, which serves as the bottom of the oceanic food chain. The tiny shells of many forams become grains of sand.

oceans; there are ten times that number if you count extinct species. Some are planktonic, meaning they drift with the currents at the water's surface, and some are benthic, meaning they live on the ocean floor. In some places, benthic forams are so dense that the ocean sediment is made almost totally of their microscopic shells. Called tests, these tiny shells are made mainly of calcium carbonate. They catch food using a network of thin cytoplasmic extensions that project through tiny holes in their shells.

The tests of forams look delicate, yet they are often found in near-perfect condition when viewed through the microscope. Forams, which can create beautiful shapes, vary in size and color from region to region. This partly explains the variations in texture and color of some beaches. The pink sands of Bermuda owe their color to a foram called *Homotrema rubrum*, which has pinkish-red tests. The red coloration is due to an iron salt incorporated into the skeleton.

Something else is incorporated into these tiny skeletons that is essential for the health and balance of our planet: carbon. All the shells produced by microscopic marine organisms take carbon atoms from the air and the water and fix them into their skeletons, usually in the form of calcium carbonate. These tiny, shelled organisms are essential for the natural balance of the planet.

Microscopic marine organisms play a major role in Earth's carbon cycle. As humans continue to release ever-larger amounts of greenhouse gases (carbon dioxide and carbon monoxide), the atmosphere becomes oversaturated with carbon. If the carbon cannot be taken up and recycled, global warming results. Plants take up much of this carbon dioxide from the air. They utilize it to make food by photosynthesis, the miraculous process whereby glucose is made from water and carbon dioxide. Plants incorporate carbon from the atmosphere into sugar, while microscopic marine organisms incorporate carbon into their shells and skeletons. The health of our oceans is more important now than ever, because we need these organisms to complete the carbon cycle.

Coral reefs play a major role in the carbon cycle as well as in the formation of beach sand. Reefs typically determine the contents of beach sand in tropical environments. The calcium carbonate reefs provide a

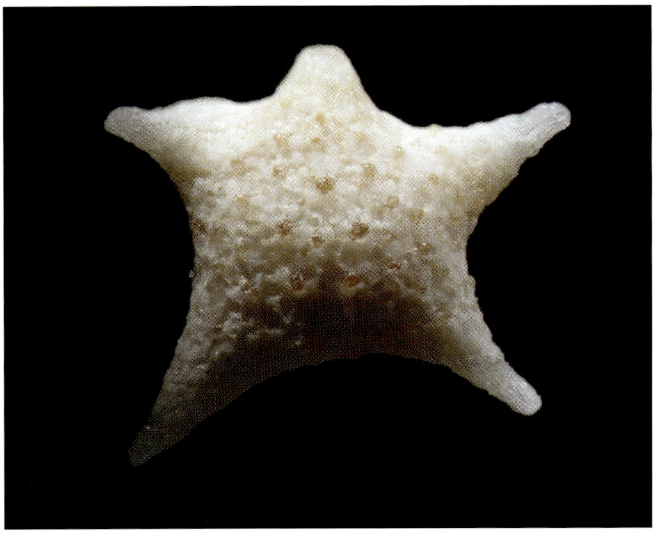

Tiny star-shaped forams are a major component of beach sand on Okinawa, Japan (magnification 50x).

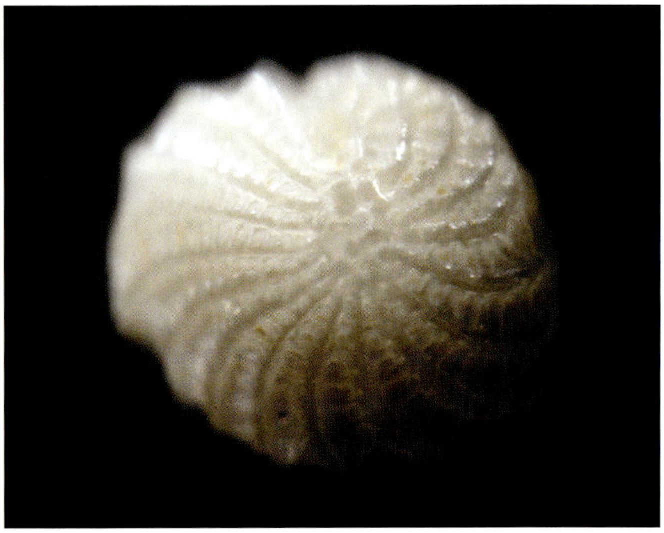

A foram from Lahinch, Ireland, has a pinwheel shape (magnification 90x).

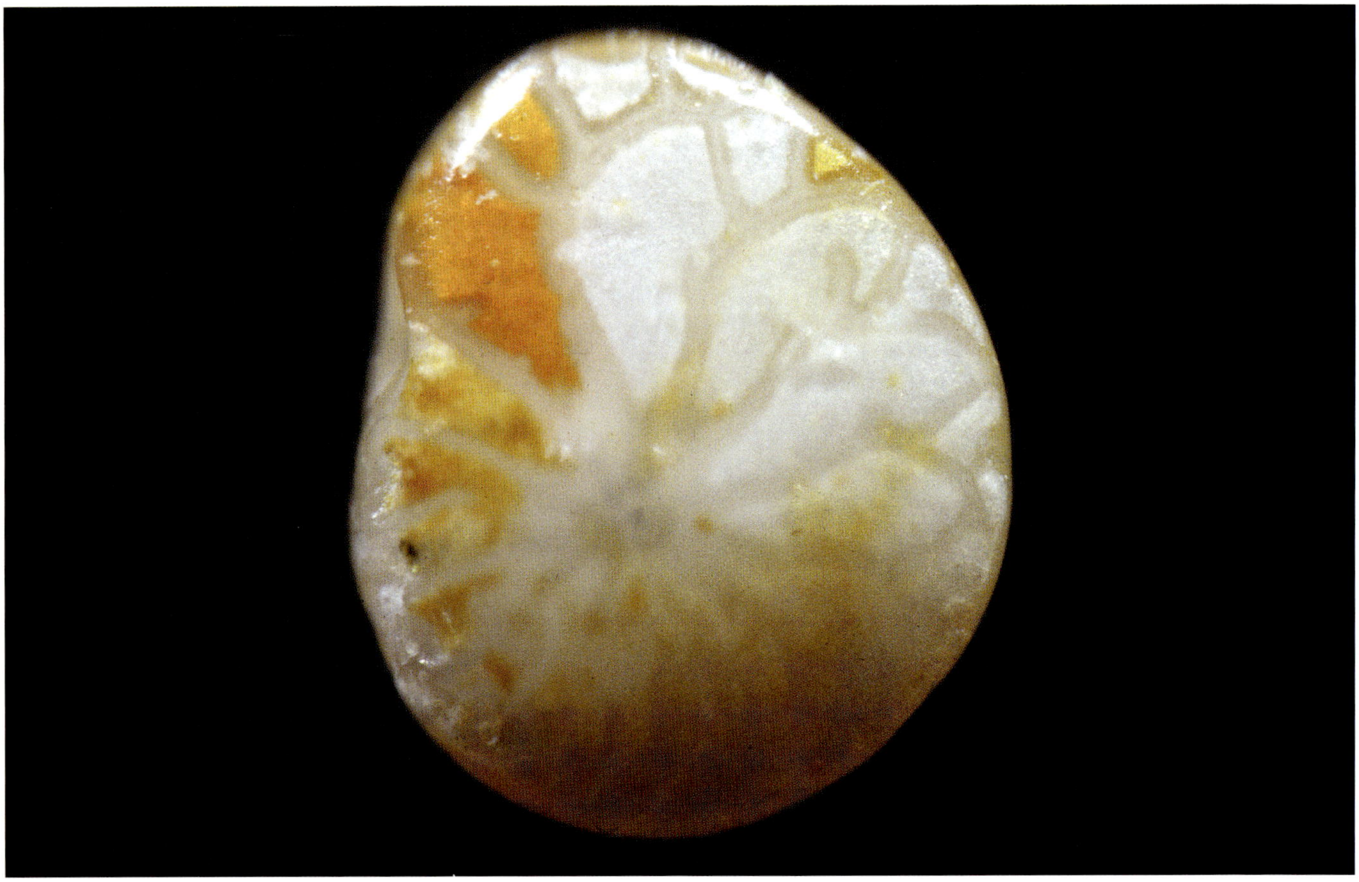

This species of foram found on some Hawaiian beaches is normally white, but this one has fossilized into a unique orange-colored grain of sand (magnification 105x).

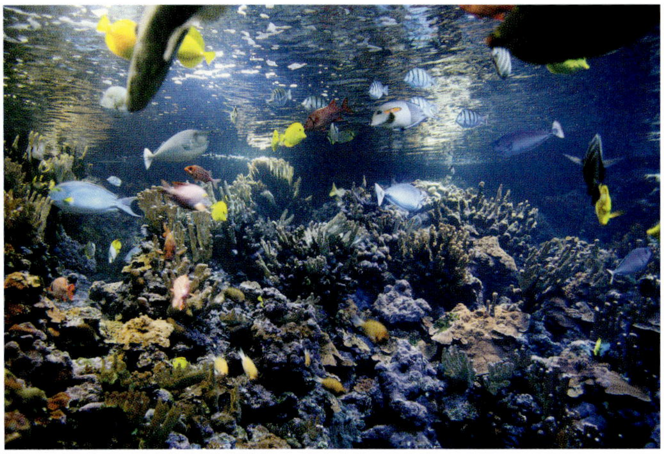

Coral reefs are a source of biological diversity and represent a major niche for ocean life throughout the world.

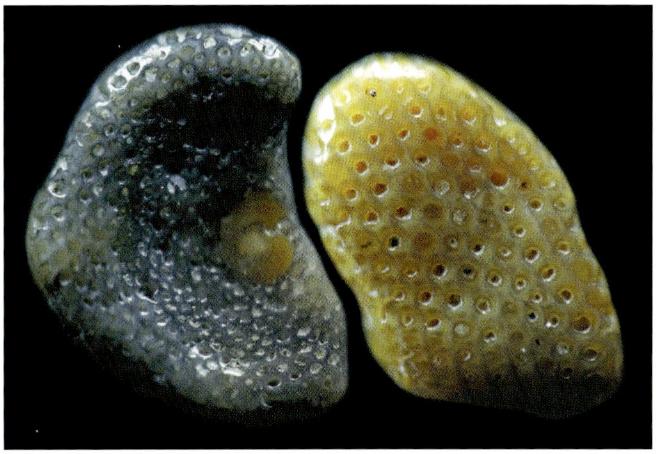

A pair of coral fragments collected from a beach in Maui. Colorful grains of coral constitute much of the sand on tropical beaches (magnification 60x).

habitat for a large variety of marine life, which is reflected in the bits and pieces that make their way to nearby beaches. The life forms that inhabit the reefs sequester carbon dioxide from the atmosphere and recycle it into their calcium carbonate shells. Human intervention has significantly reduced the number of these marine organisms, causing the levels of atmospheric carbon dioxide to increase and global warming to worsen. This is taking place at an increasing speed, and the health of coral reefs throughout the world is decreasing at an alarming rate.

When you look closely at the surface of live coral, you can see that it is covered with a myriad of tiny polyps. Each polyp has tentacles and a mouth that feeds on plankton, the microscopic plants and animals that float throughout the world's oceans. The coral produces an exoskeleton of calcium carbonate, which forms around the base of each polyp, allowing them to retreat inside for protection. The vibrant color of coral is due to the presence of algae called zooxanthellae. Different species of zooxanthellae give the coral different colors. The algae provide the coral with energy through photosynthesis and aid in calcification of the reef. In return, the coral provides the zooxanthellae a safe environment in which to flourish. This is an excellent example of a symbiotic relationship, where two different species live together for the purpose of mutual benefit.

On some coral reefs, parrotfish add significantly to the formation of sand by biting off chunks of coral while grazing on zooxanthellae. Special bones in the fish's throat grind the coral into fine grains to extract the algae. The parrotfish digest the algae and then poop out the remains, which become little bits of coral sand grains that wash up onto the beach. A single adult parrotfish can produce as much as a ton of sand per year. Microscopic bits of beach coral are particularly beautiful under the microscope.

Sea urchin spines are some of the more interesting components of beach sand. These spines break off the body of the sea urchin and then fracture into fascinating grains of sand with intricate internal structures. They can be green, white, brown, or even bright purple. The tips of the sea urchin spines look like tiny 3-D mandalas when seen in cross section. The spine is made of a single large crystal of calcite, with an intricate internal structure. Scientists are currently studying how these structures self-assemble into such elaborate forms. The future of nanotechnology will be guided by the knowledge gained from discovering how different life forms control the progression of biomineralization, the process by which bones, teeth, and shells are formed.

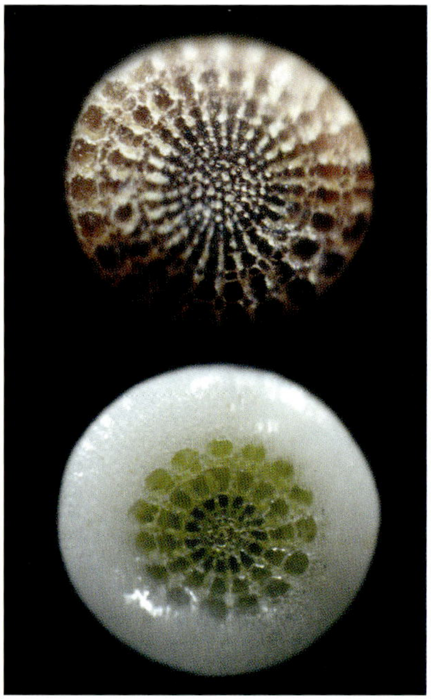 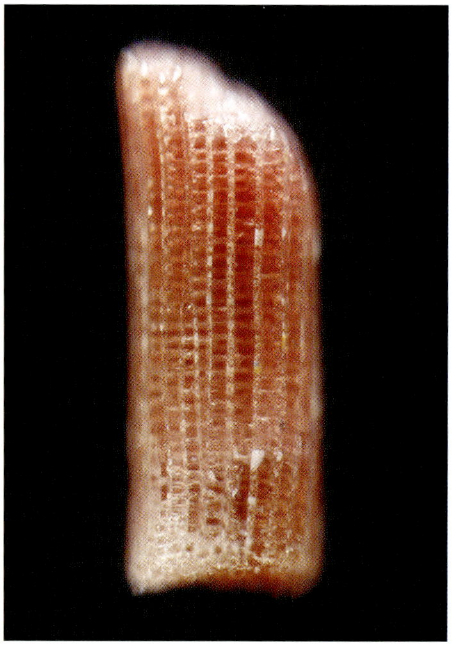 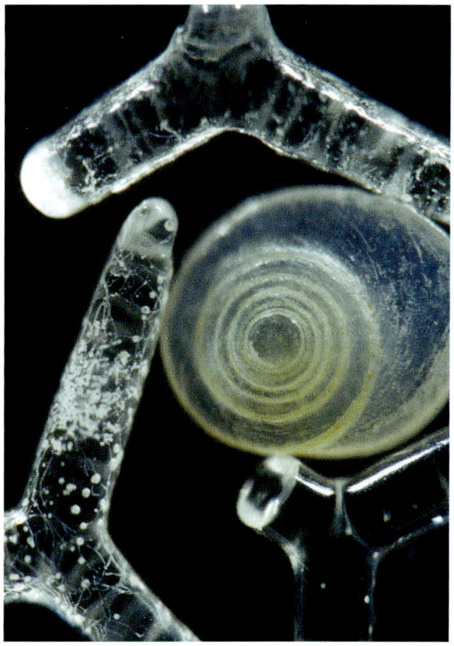

Left: Fragments of sea urchin spines, when viewed in cross section, look like mandalas. The tips have broken off and been deposited as grains of sand on a Maui beach (magnification 80x). *Middle:* A fragment of a sea urchin spine shows its intricate internal design in side view in sand from Maui (magnification 80x). *Right:* Sand grains in the form of glasslike needles are actually microscopic sponge spicules, the internal skeletons of sponges. The spicules surround the tip of a spiral shell that has broken off and eroded into a beautiful, translucent grain of sand (magnification 70x).

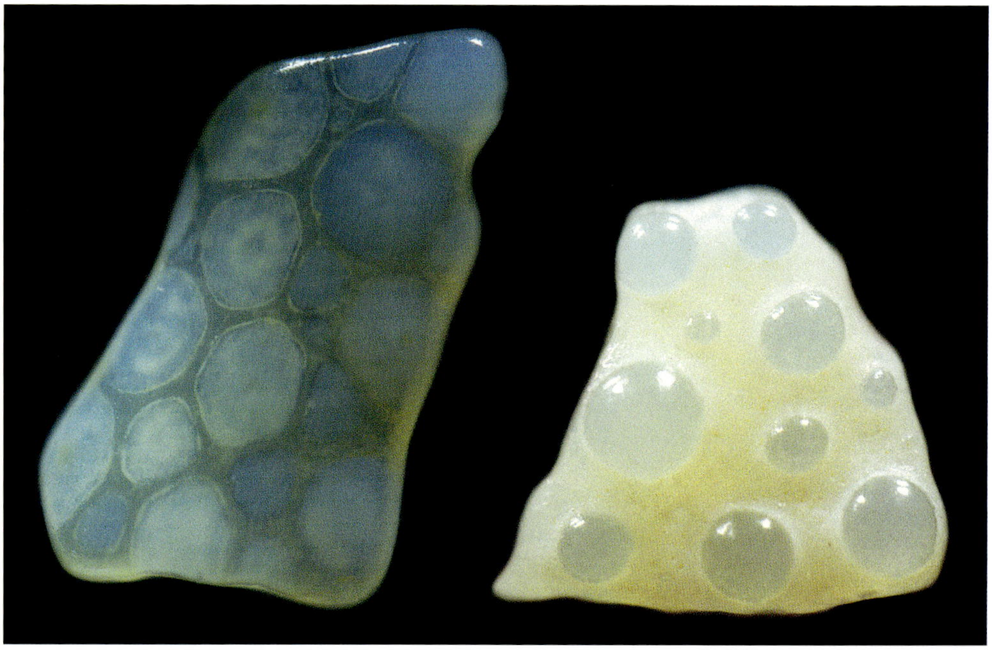

Two grains of sand are decorated with interesting circular and spherical structures. They are tiny fragments of a baby sea urchin shell. The raised bumps on the white grain represent the site of insertion for the sea urchin's spines. The blue grain has eroded to the point that the raised bumps have been completely rubbed off (magnification 100x).

Another fascinating element in sand is the sponge spicule. Sponges contain thousands of microscopic glasslike needles called spicules that function as an internal skeleton. These spicules come in many different shapes, depending on the species of sponge. They are often found as triple radiating needles, but they can also take on shapes that resemble plates and boat anchors. Some sponge spicules are made of calcium carbonate or silica, while others are made of a substance called spongin. The spongin-type sponges were the ones once used as scrubbers in bathrooms before being replaced by the synthetic spongelike material used today. In contrast, the silicate-type spicules are very sharp, and touching them can be harmful. An acquaintance of mine recently went snorkeling and spent some time "playing" with the sponges by squeezing them between her fingers. When she got out of the water, she noticed her fingers were red and swollen, and she ended up with a severe staph infection. Warm tropical waters contain a host of different bacteria, including *Staphylococcus aureus*, which was injected into her fingers by the needle-sharp sponge spicules.

Tiny bits of broken seashells add color to the sand grains on many beaches. Mollusks spend their lives buried in the sand as protection against the surf. The inside of a mollusk shell is often made of mother-of-pearl, which is a form of calcium carbonate. Some white-sand beaches in tropical regions are made up almost entirely of bits of broken seashells. The coloration of shells adds to the peppered appearance of sandy beaches. An iridescent bit of broken mother-of-pearl can become a beautifully unique grain of sand.

Precipitated Components

Some sand grains, such as ooliths, are formed from the precipitation of calcium in solution. These rounded grains form around a smaller particle that acts as a nucleation site for calcification.

Even some types of seaweed can become of a source of sand. Decaying calcium-bearing seaweeds, like

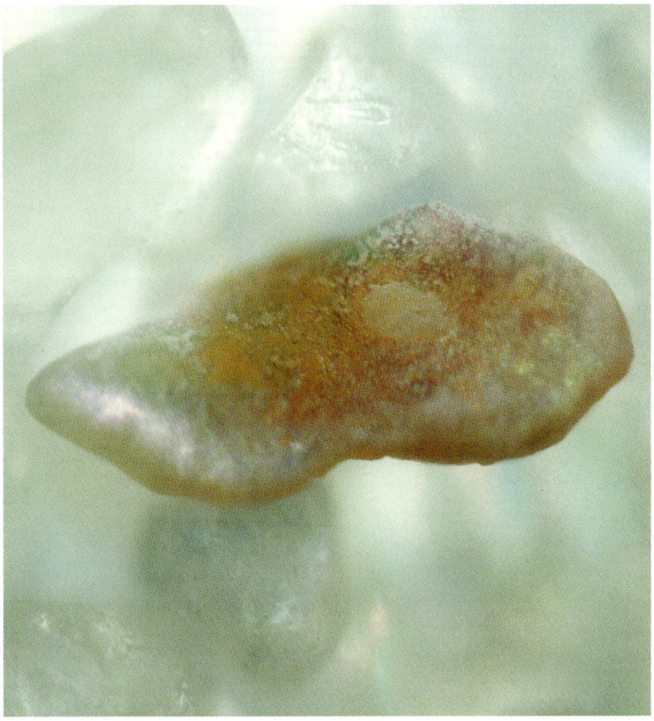

An iridescent grain of sand was collected from Australia. It's a little fragment of mother-of-pearl (magnification 110x).

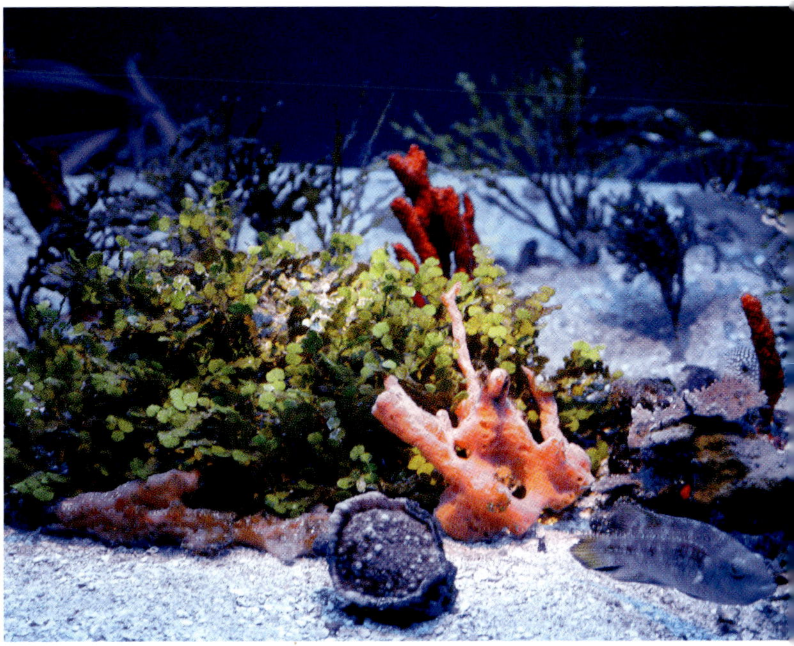

Halimeda, known as the "money plant," can calcify and turn into grains of sand.

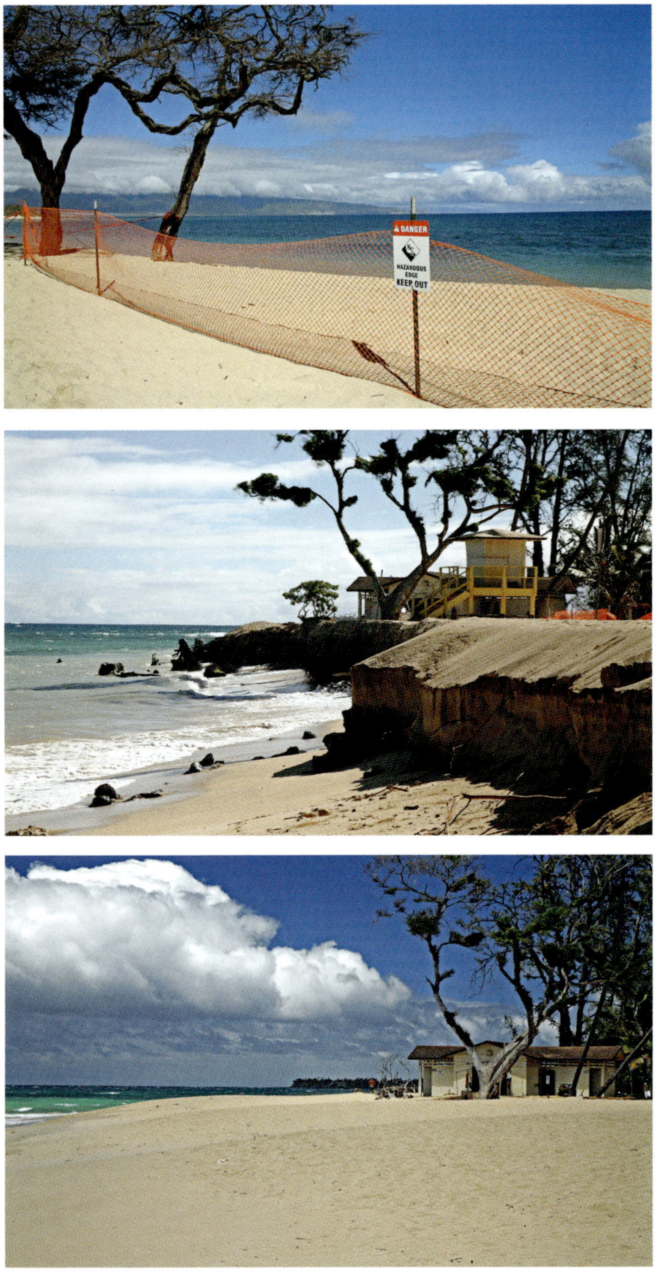

Baldwin Beach, Maui, is a dramatic reminder that sand is a dynamic landform. Powerful winter waves removed most of the sand from the beach in a couple of days during a 2006 storm. *Top*: A danger sign is posted after the storm. *Middle*: The storm laid bare tree roots and encroached on a lifeguard's house. *Bottom*: A year later, much of the sand had returned to the beach.

halimeda, are important creators of some tropical sand beaches. Halimeda is sometimes called the "money plant" because it consists of flat, round segments that resemble coins. On some coral atolls, these seaweeds are responsible for as much as 90 percent of the sand.

The Movement of Sand

The sand on a beach appears stationary to the casual observer, but beaches are actually dynamic landforms in constant change. Waves, currents, and winds drive a continual process of creation and erosion. In order for the size and shape of a beach to look the same week after week, the rate at which sand is being added must equal the rate at which it is being eroded. On a macroscopic scale, the entire beach undergoes continual change, and even on a microscopic scale, the individual sand grains erode and change shape as well.

Wave action moves huge amounts of sand daily. As waves approach shallow water, they slow down due to friction with the ocean floor. The waves get steeper because the bottom of each wave moves more slowly than the top of the wave, and finally the waves break. During the slowing and breaking process, sand is transported up the beach; during the backwash of the wave, sand is moved back down the beach toward the ocean. The movement of the sand takes a zigzag motion. This phenomenon occurs because waves usually approach the shore at an oblique angle, moving the sand at that angle as the wave hits the beach. When the wave retreats, gravity pulls the water straight back down the beach, resulting in the zigzag movement. The size and power of the waves affect the rate of transport of sand. The transport of sediment along the seashore is part of a process called littoral drift.

When you stand on the beach with your feet in the water, you feel the frantic race of sand rushing past your feet. The movement of the sand is so powerful that it literally undermines your footing on the shoreline. If you put on goggles and swim out just past the wave break, you can look down at the sand and watch it move with each successive wave.

Seasonal cycles of sand deposition and erosion can dramatically affect the appearance of beaches from season to season. Beaches can be wide and gently sloping in the summer, then become steep and narrow in winter. They can literally disappear overnight, stripped naked by powerful waves. The movement of sand is truly dramatic when driven by hurricane-force winds. Much of the sand removed from beaches during winter storms is deposited on sandbars just offshore, and it is often returned to the beach by moderate summer swells.

Strangely enough, during the writing of this section of the book, I drove to my local beach for a morning swim and was confronted with a bright-orange cyclone fence holding a printed notice that read "Danger – Hazardous Edge – Keep Out." Walking around the fence to investigate, I could see that the entire beach had been wiped out in a recent storm. I was stunned. The lifeguard's house, which used to be about 50 meters from the waterline, was almost consumed by the waves. Tree trunks from the previously expansive beach were snapped and the roots laid bare by the raging surf. A year later, a large amount of the sand had returned, but not as much as had been there previously.

In Florida, the situation is worse. Storms brewing off the warm Atlantic waters in the summer can turn into hurricanes. They can sweep away miles of Florida beach in a matter of hours. Coastal engineers try to rebuild the beaches by mining sand from offshore and depositing it on the beaches. But global warming is making their job harder, as hurricanes are becoming more intense and the ocean level is rising in response to polar ice melting to liquid.

Florida is running out of sand, which is a valuable natural resource that fuels a huge tourist industry. A new experimental program takes an interesting approach. It recycles bottles and crushes them into sandlike particles. Tons of the ground glass is distributed back onto the beaches to replenish the sand. It's a fascinating solution, considering glass originally came from sand.

At the other environmental extreme, the movement of sand results in the construction of thriving beach ecosystems in the form of extraordinary sand dunes. Dunes form only where there is sufficient wind and a constant source of sand. The growth of dunes begins when blades of grass, a stone, a fence, or some other object obstructs the wind across an open beach. The dunes grow grain by grain and gradually move inland. The wind usually creates a gentle slope along the face of the dune; sand carried over the crest falls abruptly behind the dune, resulting in a steep slope on the backside.

The Ecology of Beaches

The coastal saltwater environment forms the foundation of one of Earth's most important ecological domains. Beaches fill distinct environmental niches inhabited by a variety of different plants and animals.

The *intertidal zone*, located between the highest high-tide mark and the lowest low-tide mark, is an especially harsh environment that varies between extremes. It often contains trapped areas of seawater, creating tide pools with high salinity. Each and every day, tidal zones go from baking in the hot sun to submersion under seawater and back to cracking hot sun again. Added to that is the relentless crashing of

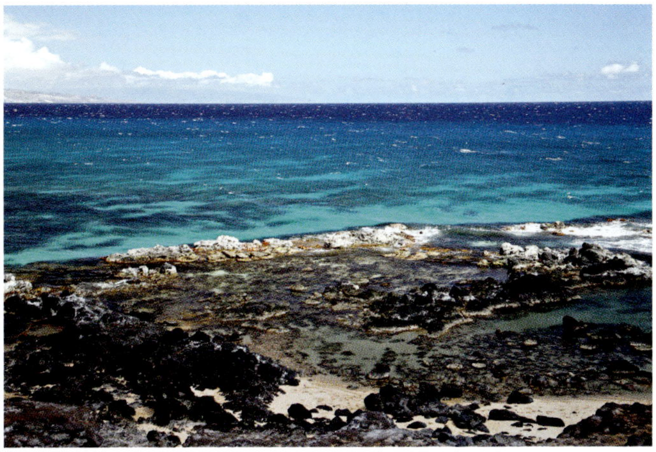

The intertidal zone is a harsh and ever-changing location, yet some forms of life thrive in this turbulent environment.

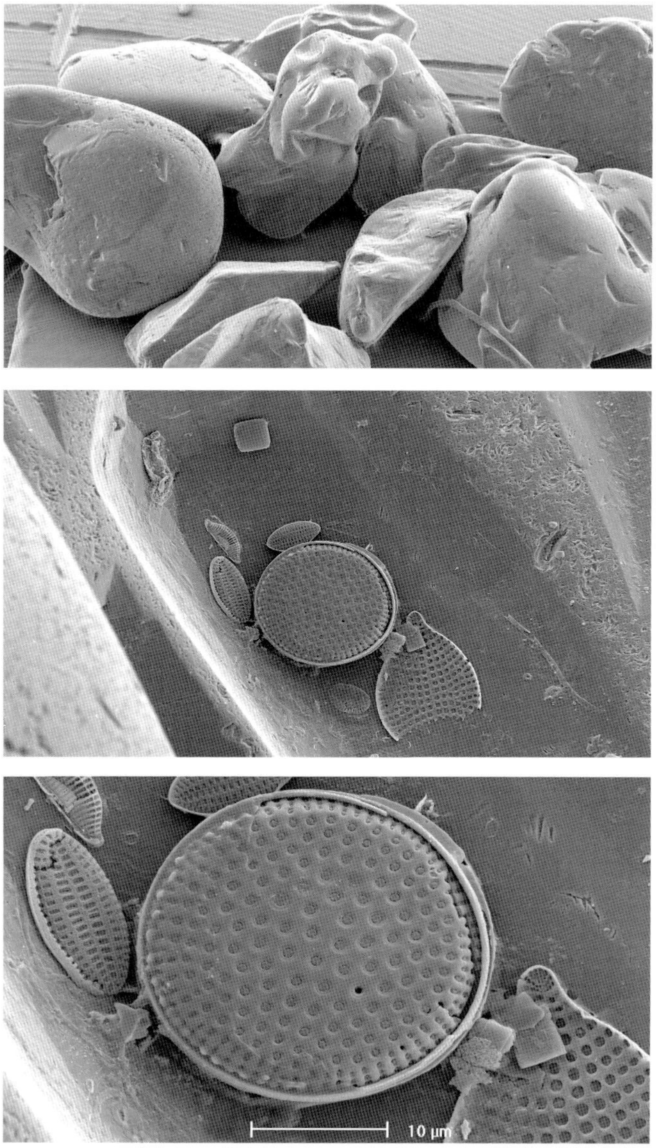

A scanning electron microscope shows diatoms adorning the surface of quartz sand grains from Ipanema beach in Rio de Janeiro, Brazil. *Top*: The sand grain on the far right harbors a handful of diatoms in a tiny indentation (magnification 100x). *Middle*: A close-up view of the indentation in the sand grain reveals the group of diatoms (magnification 420x). *Bottom*: A close-up shot of the diatoms shows the tiny silicate shells (magnification 800x). *Dr. Aviva Burnstock*

the waves. Yet, many animals and plants thrive in the intertidal zone.

The intertidal zone is inhabited by a multiplicity of invertebrates and insects such as bivalve mollusks, crustaceans, sea urchins, starfish, sponges, and tube-building worms. They have acclimatized to the harsh environment of tide cycles, buffeting waves, high salinity, and predators. When these animals die, they leave behind their skeletons and shells, which are gradually ground into beautiful grains of sand by the action of the surf.

Some intertidal species burrow into the sand to protect themselves. Oysters, clams, and other bivalve mollusks are burrowers. At low tide, water retained between the sand grains is filled with billions of microscopic diatoms and zooplankton, upon which buried bivalves feed. The mollusks have long siphons with fine screens that filter out sand particles and allow water and organic material to pass through, providing a plentiful food supply.

A host of organisms thrive near the surface of the sand at low tide, the sand remaining wet due to capillary action. Capillary action is the movement of water within small porous spaces due to the forces of surface tension and adhesion. The spaces between the sand grains create fine passageways for water to move by capillary action. The organisms vary from tiny bacteria to worms and small crustaceans. The bodies of these animals are thin and long, and they can move easily among the grains. They have adhesive organs that allow them to stick to the sand grains and not be washed away. Great quantities of diatoms, forams, dinoflagellates, protozoa, annelids, and nematodes live in the wet sand.

A few years ago, Dr. Aviva Burnstock, reader lecturer at the Courtauld Institute in London, brought home sand from a visit to Brazil. She looked at the quartz sand grains through a scanning electron microscope. To her surprise, the sand was decorated with tiny diatoms. Diatoms are a type of algae that create beautiful silicate houses. One of the major forms of plankton, the diatom represents a significant portion of all ocean life. Each species makes a microscopic shell with its own particular shape and design; each one is more beautiful and delicate than the next. Diatoms are wonderful examples of worlds within worlds in a grain of sand.

The *subtidal zone* extends from the lowest part of the intertidal zone to as far out in the ocean as sand is moved by wave action. Environmental factors change slowly there, and animals such as eels, flounders, soles, rays, and skates have adapted a burrowing lifestyle.

The highest place where the ocean washes onto the beach is called the strand line. This area collects stranded material, which provides a cool, moist environment for small invertebrates like kelp flies, biting fleas, centipedes, beetles, and amphipods. Gulls and sandpipers scavenge for organic materials at the strand line, and bacteria break down the remainder, releasing nutrients back into the ocean, completing the cycle of life.

The upper beach, or *supratidal zone*, ranges from the high-tide mark to the sand dunes. Here, animals live under the sand to avoid exposure to the sun-baked surface. The upper-beach sand contains almost no food or water. The most common animals in this zone are sand fleas and crabs, which take refuge under driftwood and seaweed during the day. Both are crustaceans that consume dead plant and animal material.

Sand dunes create an interesting ecological environment. Frontal dunes closest to the water often host a plant and animal community. The specialized plants have adaptations such as waxy coverings on the leaves, small leaves with few stomata, large root systems, and thick stems and leaves to store water. The first line of dunes is the primary dune, which deflects ocean breezes and creates a semiprotected environment on the backside of the dune. Plants that can't withstand direct hits of salty wind grow in the sheltered side of the primary dune. The second line of dunes is often thickly vegetated, behind which a coastal jungle frequently develops.

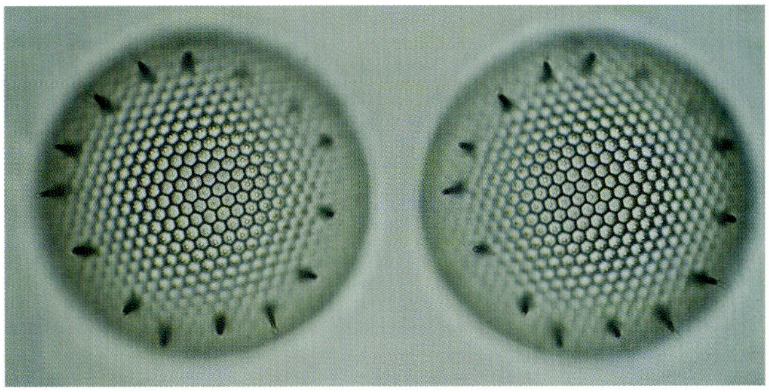

A stereo image of a tiny diatom looks like geodesic domes. If you cross your eyes, you can see a 3-D image of the beautiful diatom (magnification 350x).

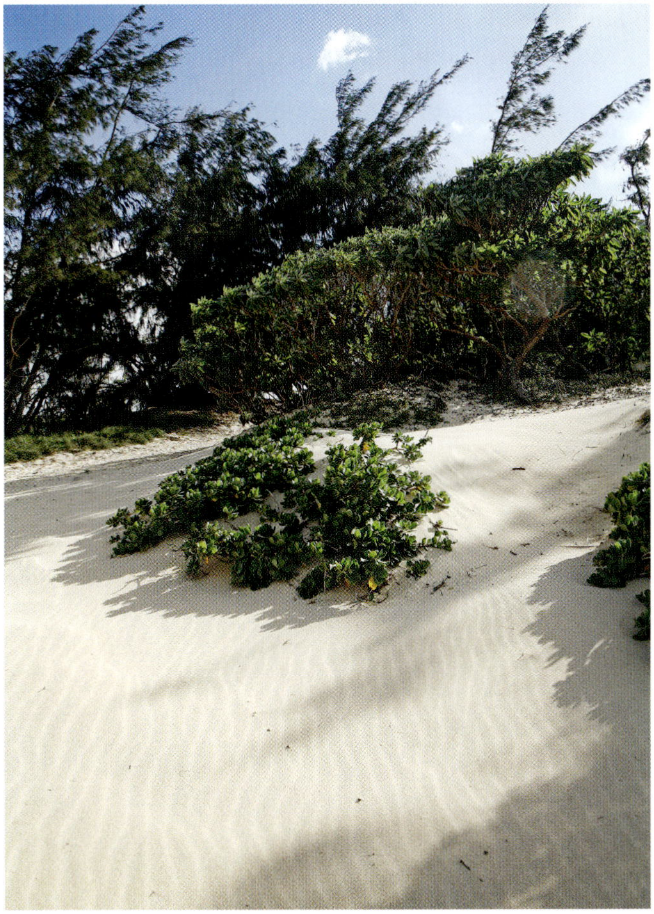

Sand dunes form with the right combination of wind, sand, and plants or other obstructions on the beach. The pattern of waves in the sand is evidence that the wind has blown the sand up the beach.

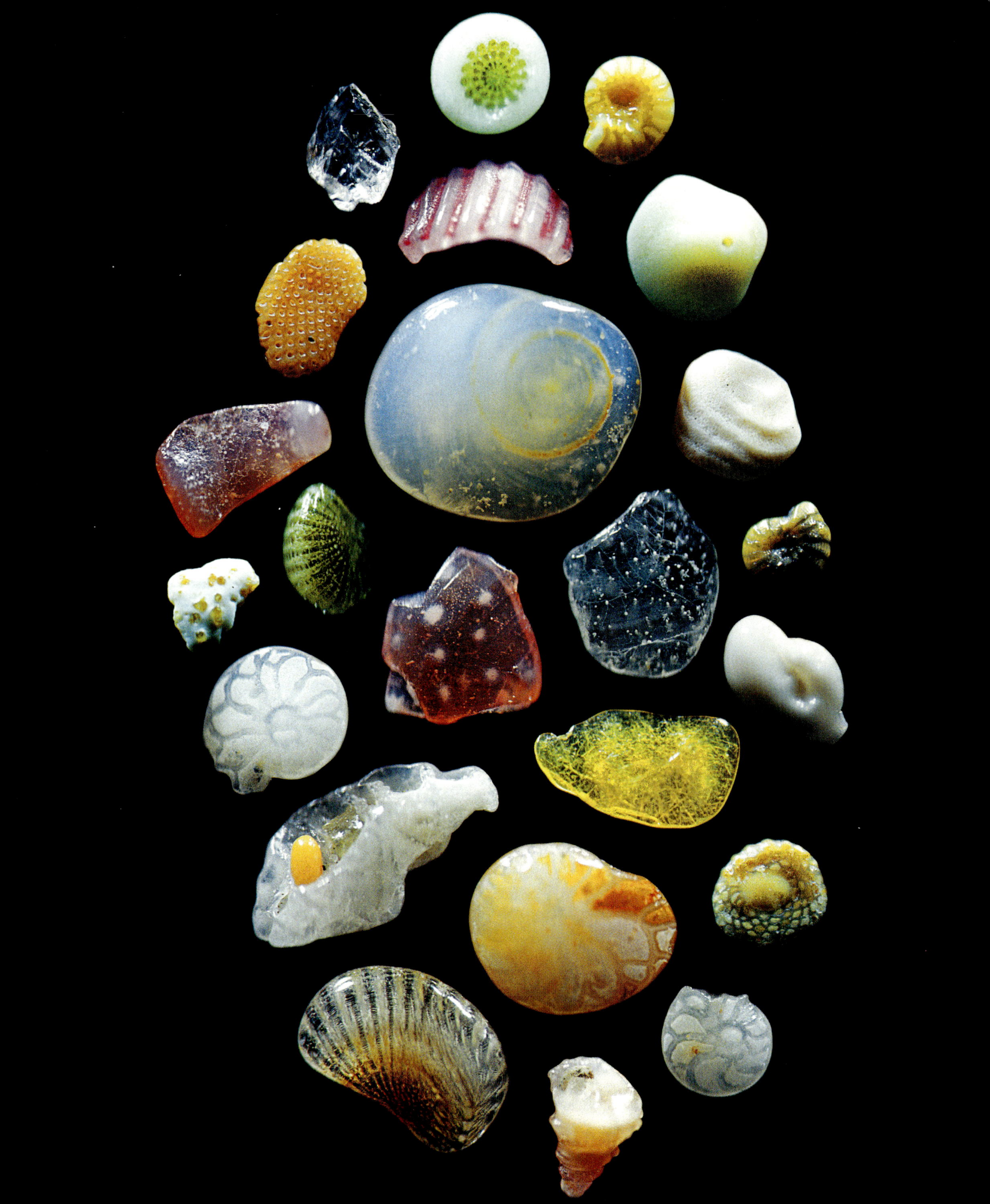

CHAPTER 4

Colors in Sand

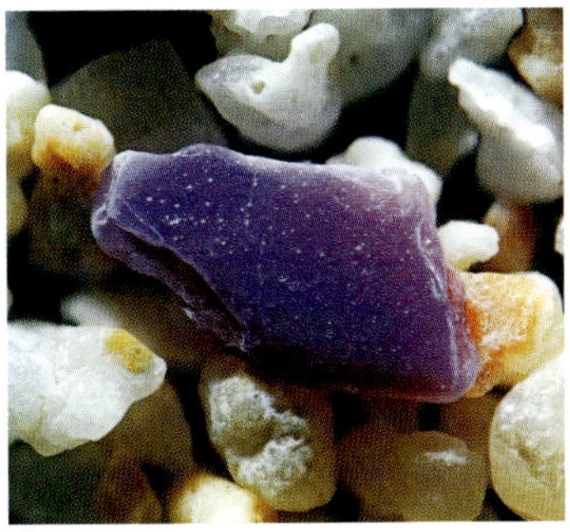
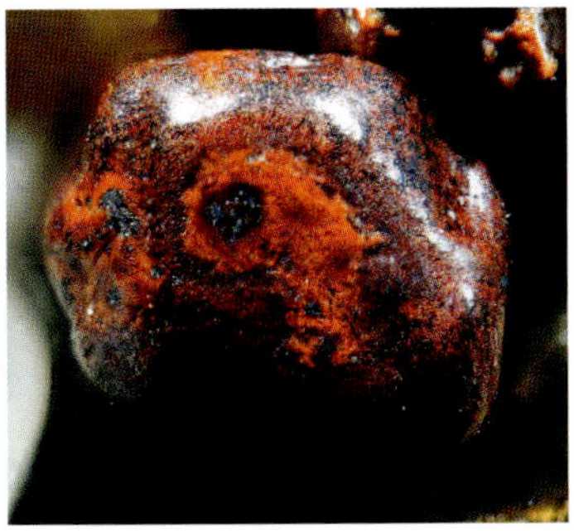
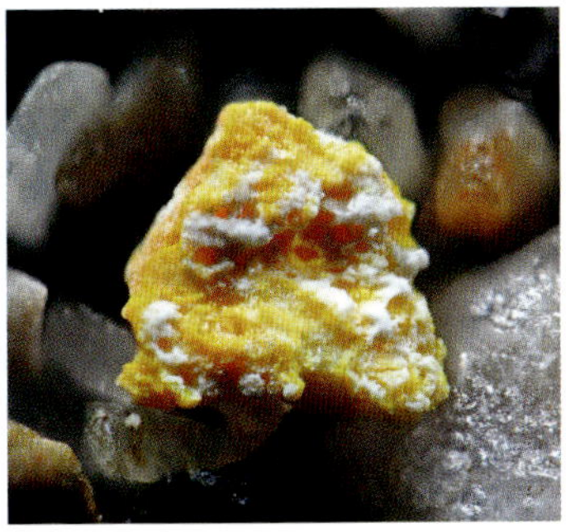

Although most beaches take on a single muted color to the naked eye, sand is full of diverse colors when looked at closely. The medley of images presented in this chapter illustrates the range of dramatically different colors in sand grains from around the world.

Sand is not just a bunch of small, round, beige-colored stones. In fact, sand grains come in every color imaginable. The bright pinks, reds, and greens of mineral sand are stunning, very much like gemstones except extremely small. In fact, some of them actually *are* tiny gemstones. These little bits of quartz, amethyst, olivine, and garnet catch the light and bounce it around like sparkling jewels.

Biological components also sprinkle sand with fabulously bright colors. For example, sea urchin spines and coral can be found in a wonderful range of beautiful colors.

Most of the colors seen in the sand grains in this book are natural. But that statement begs the question, what do I mean by "natural"? The colors in this book are natural in that they have not been manipulated or changed. But sand looks different

The spectacular natural colors of sand are seen in an arrangement of grains from a beach in Maui. The sand contains colorful shell fragments, coral, sea urchin spines, forams, and tiny intact shells. The sand grains reflect the diverse biological life forms found in tropical waters (magnification 100x).

depending on how you view it. From a distance, sand on a beach may appear light brown because all the colors merge. When we look at the beach more closely, we notice that it is peppered with individual sand grains of various colors. And through a microscope, the individual grains look entirely different. They even vary in appearance depending on the magnification level, the lighting conditions, and what instruments you use.

Instances of the colors being "not natural" can be seen in images where I've employed crossed polarizing filters. Polarizing filters only pass light that is vibrating in one particular plane (such as up and down) and they block light that is vibrating in all other planes (such as right and left). Two of these filters can be used in such a way that the first filter illuminates the sand grains with polarized light, and the second filter analyzes the path the light takes as it travels through the sand grains. Certain materials have the ability to rotate the plane of polarization of light, a phenomenon known as birefringence. Some sand grains, such as quartz and calcite, are strongly birefringent. Dramatic primary colors are seen by viewing in this fashion.

So are these colors "natural"? Birefringence is a phenomenon of nature, so it surely must be natural. It's just that those colors can't be seen without the polarizing filters. These filters provide us with a new set of eyes for viewing nature, a fresh point of view that provides a new dimension of information. Interestingly enough, some animals, such as bees and migratory birds, see polarized light naturally. They use this special sense to navigate from place to place, even during overcast days when the sun isn't out. In fact, it has recently been shown that migratory songbirds use polarized light to recalibrate their internal magnetic compass at sunrise and sunset.

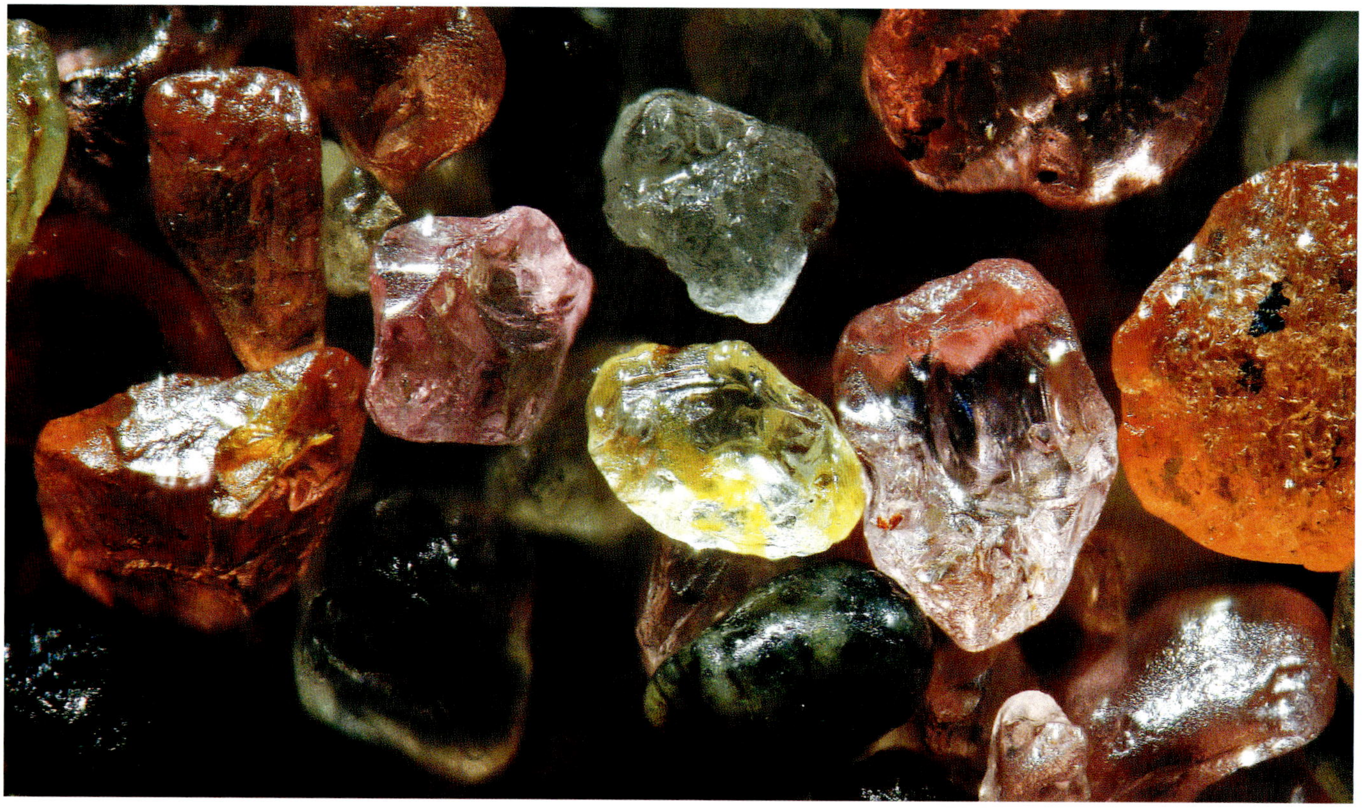

Sand from Skeleton Beach in Namibia contains rounded and polished pink-and-red garnet and angular black magnetite (magnification 170x).

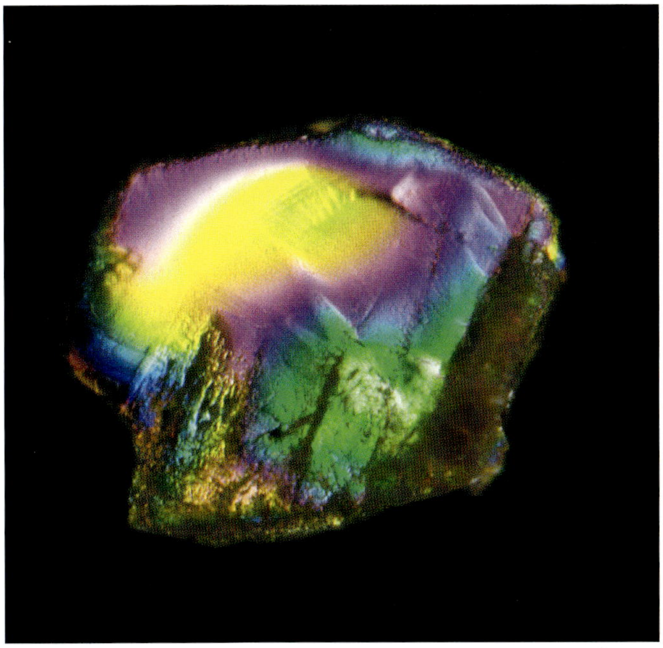

A quartz sand grain seen through crossed polarizing filters shows rainbow birefringence. The colors indicate the extent to which the crystal has rotated the plane of the polarized light (magnification 160x).

The Colors of Mineral Sand

Vibrant color is a stunning phenomenon in minerals, but mineralogists tend not to rely very much on color as an identifying characteristic. Some minerals, like garnets, come in many colors due to changes in composition or the presence of impurities. Garnets can be blood red, pale pink, brown, black, green, or orange depending on their chemistry. Metallic minerals, however, do have a definite color.

Oftentimes, the brown or reddish hue of sand grains is due to the presence of iron oxide. Many desert sand grains, including those in the northern Sahara, are coated with a thin layer of iron oxide that precipitates from the atmosphere. Iron also gives some garnets their red color. The red color in the agate fragments from Lake Winnibigoshish, Minnesota, comes from iron.

There are associations between certain minerals. For example, garnet is frequently found with magnetite. That combination can be seen in sand from Big Sur, California, and Skeleton Beach, Namibia.

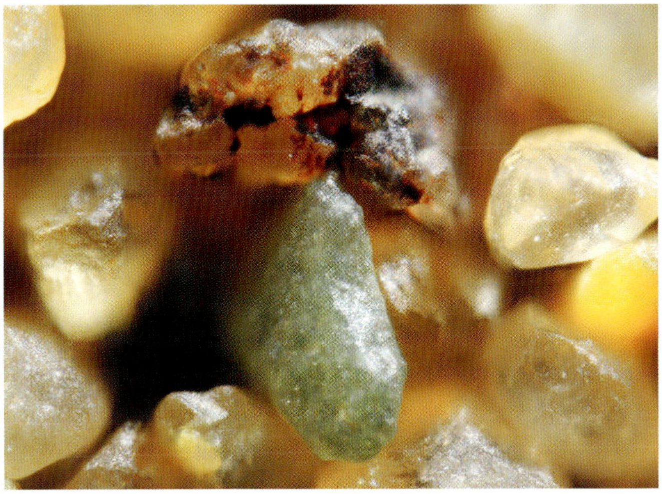

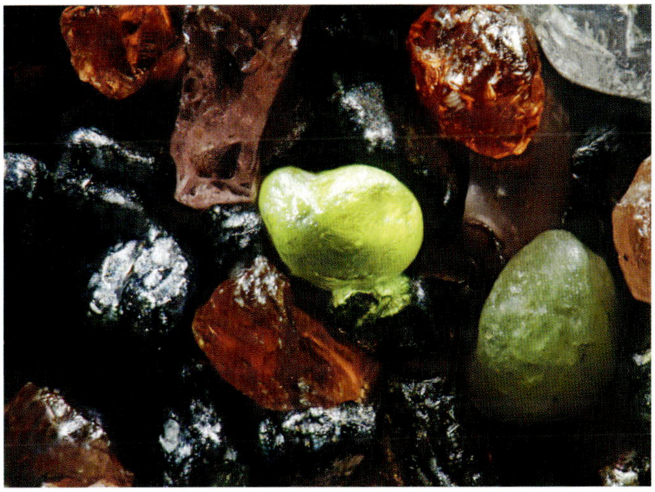

Left: Sand from the northern Sahara, Morocco, shows the typically pitted and frosted surface of desert sand, where grains constantly collide against one another. The red color of many desert sands is caused by a thin layer of iron that precipitates from the atmosphere and coats the grains (magnification 85x). *Right:* The glacially deposited sands around Lake Winnibigoshish, Minnesota, contain abundant sediments from the igneous and metamorphic minerals of the Lake Superior basin. A sample includes pink garnets, green epidote, iron-rich red agates, black magnetite, and hematite (magnification 85x).

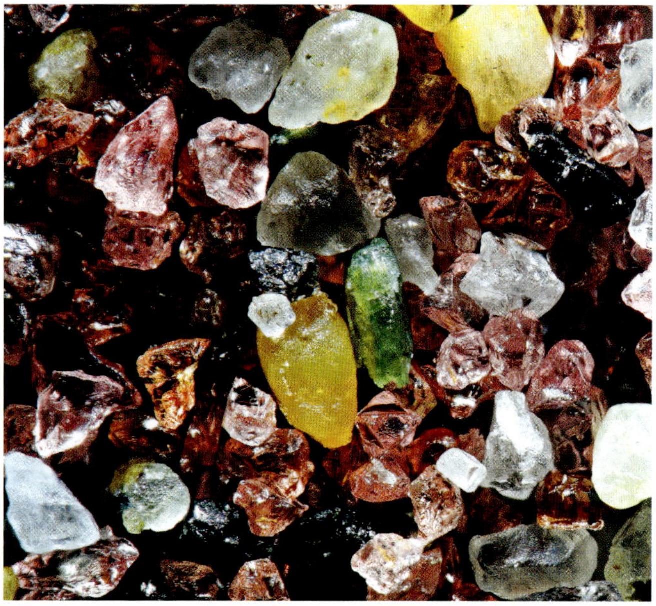

From a distance, sand from Big Sur, California, glimmers pink from garnets. Along with the garnets, there are grains of magnetite, a form a green jade called nephrite, and clear quartz. The sharp, angular grains indicate that this sand has probably moved only a short distance from its parent bedrock (magnification 45x).

Some of the most colorful sand grains are heavy mineral sands of garnet, magnetite, rutile, olivine, and epidote. Under the microscope, these gemlike sands are strikingly beautiful. Sand on the beaches at Plum Island, Massachusetts, and Lake Winnibigoshish contain some good examples of heavy mineral sands.

Heavy minerals sands can populate entire beaches or concentrate in swaths of "black sand." These mineral grains accumulate by the sorting action of the wind and by the rolling and depositing action of waves. Accumulations of heavy mineral sands often tint a beach a distinct color, such as pink, black, or green. These sands are relatively hard on the Mohs scale of mineral hardness, which ranges from talc as the softest and diamond as the hardest. Because they are hard and chemically resistant to weathering, heavy mineral sands have endured millennia of constant wave energy and transport over long distances.

Heavy minerals separate from other beach sand grains because they have a higher-than-average specific gravity (or density). Common quartz has a specific gravity of 2.65. In contrast, the specific gravity of garnet is 3.5, and magnetite is 5.1.

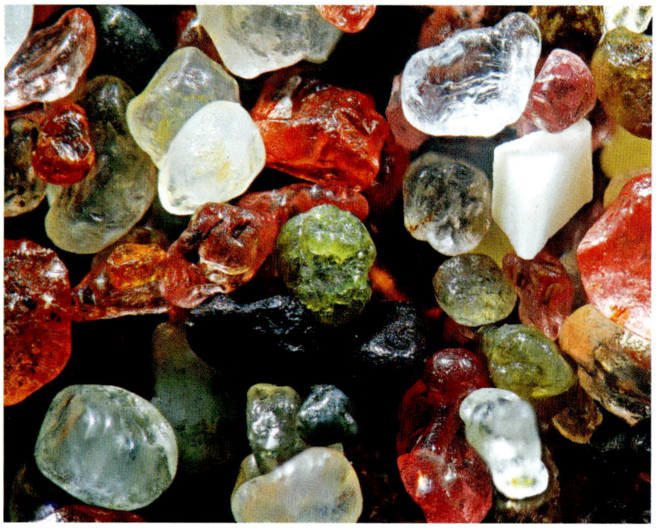

Sand from Skeleton Beach in Namibia contains rounded and polished pink-and-red garnet. The black grain is magnetite (magnification 95x).

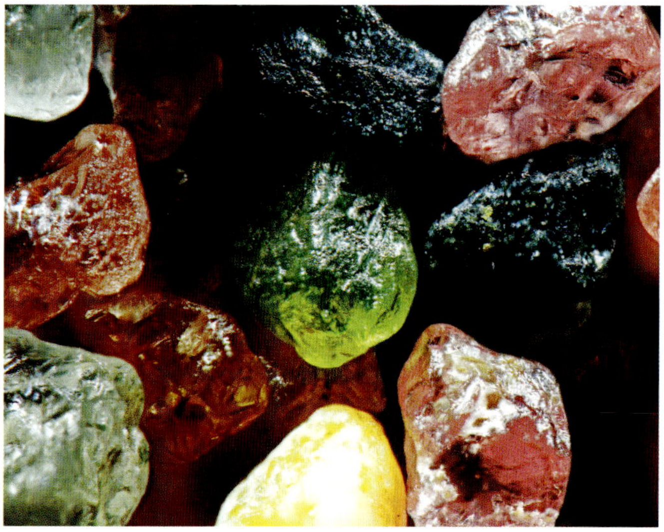

Sand from Plum Island, Rowley, Massachusetts, contains garnet, magnetite, and epidote (magnification 95x).

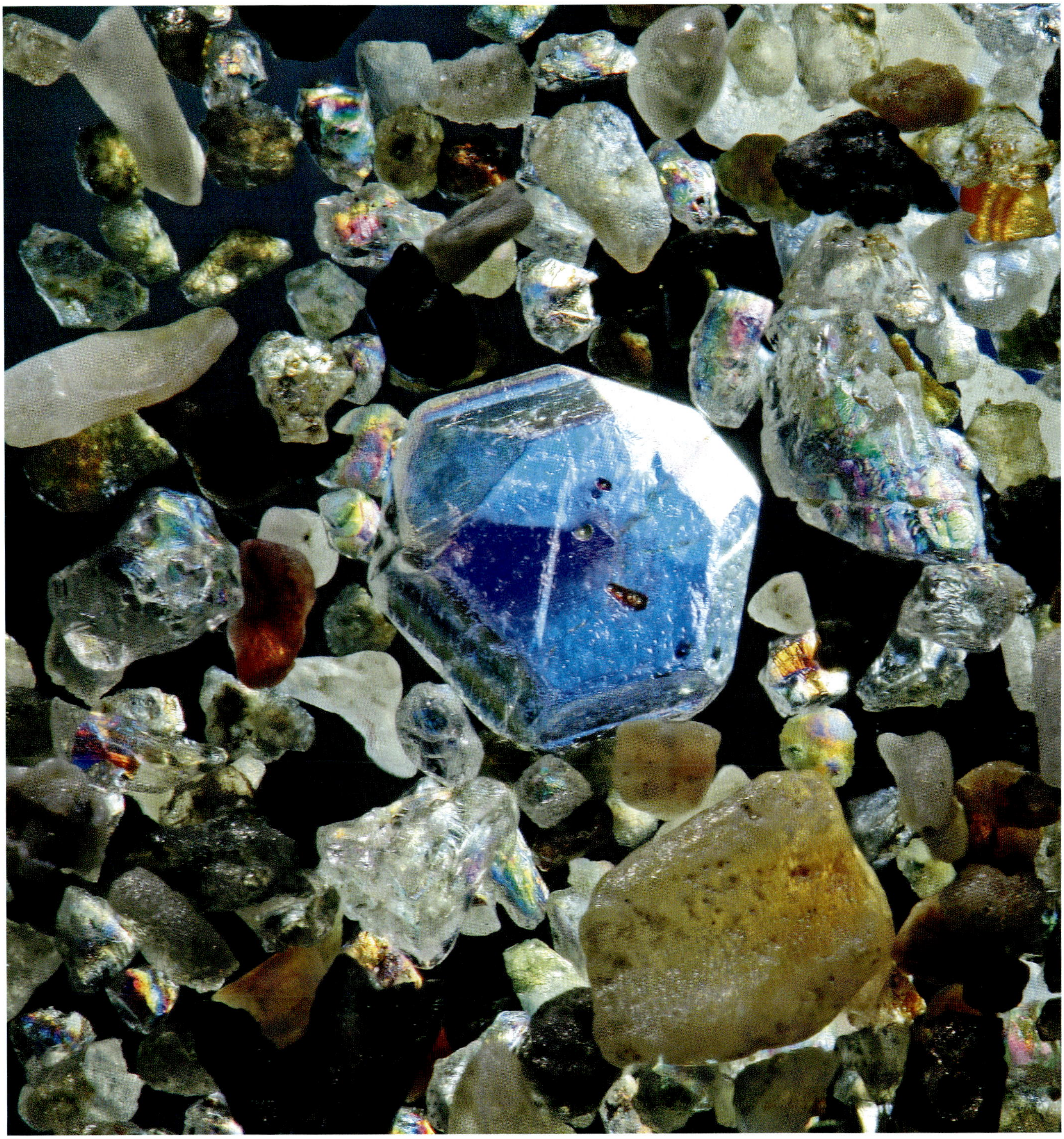

A miniature hexagonal crystal was found in sand on Zushi Beach, Japan. Crossed polarizing filters create the blue color, which indicates the orientation of the crystal structure (magnification 100x).

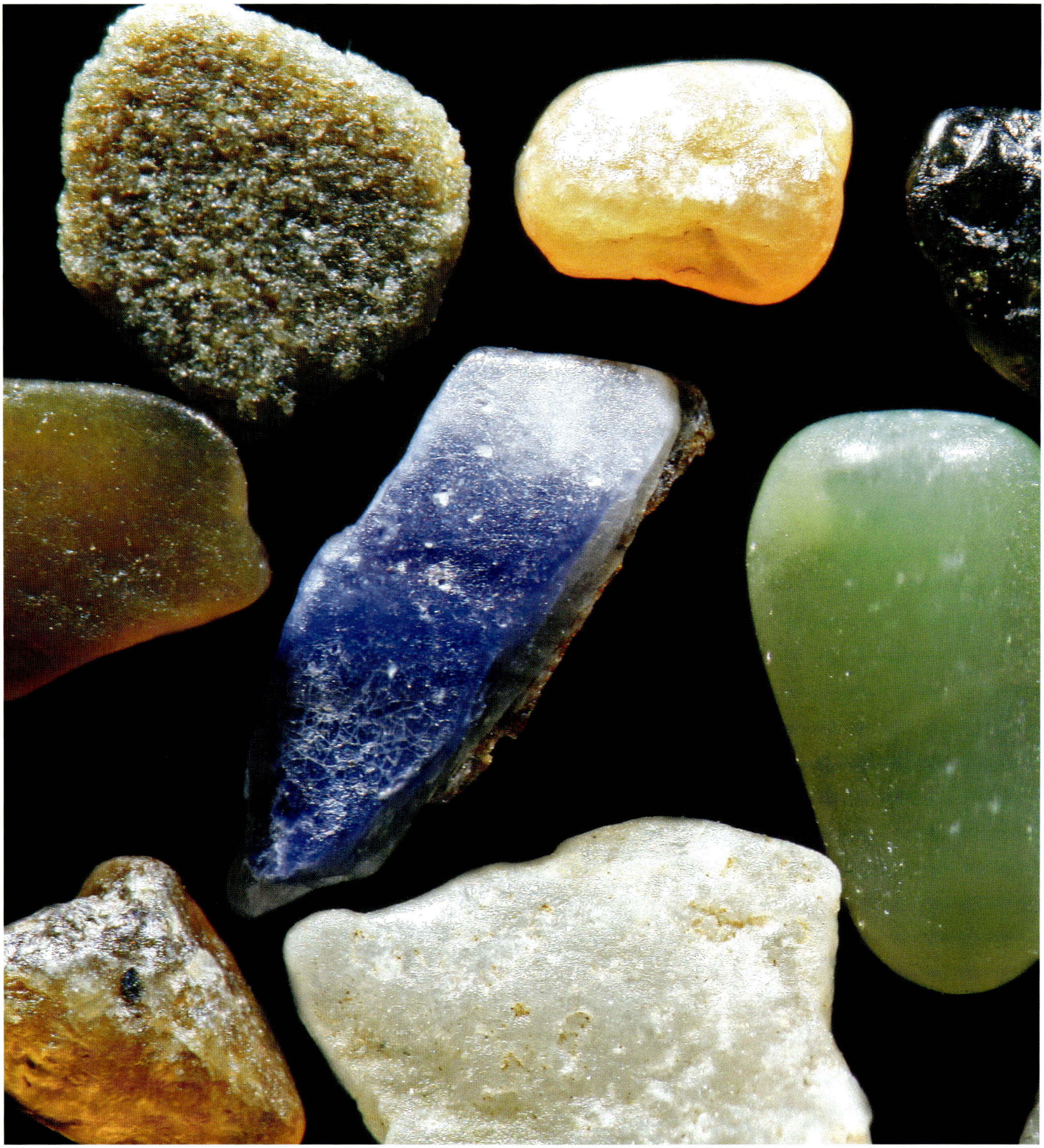

Multicolored sand from Cecina, Italy, resembles miniature gemstones. The blue fragment could be a remnant of a glazed porcelain piece from an old dinner plate or tile. The ocean has long been used as a dump, so man-made objects often become grains of sand (magnification 220x).

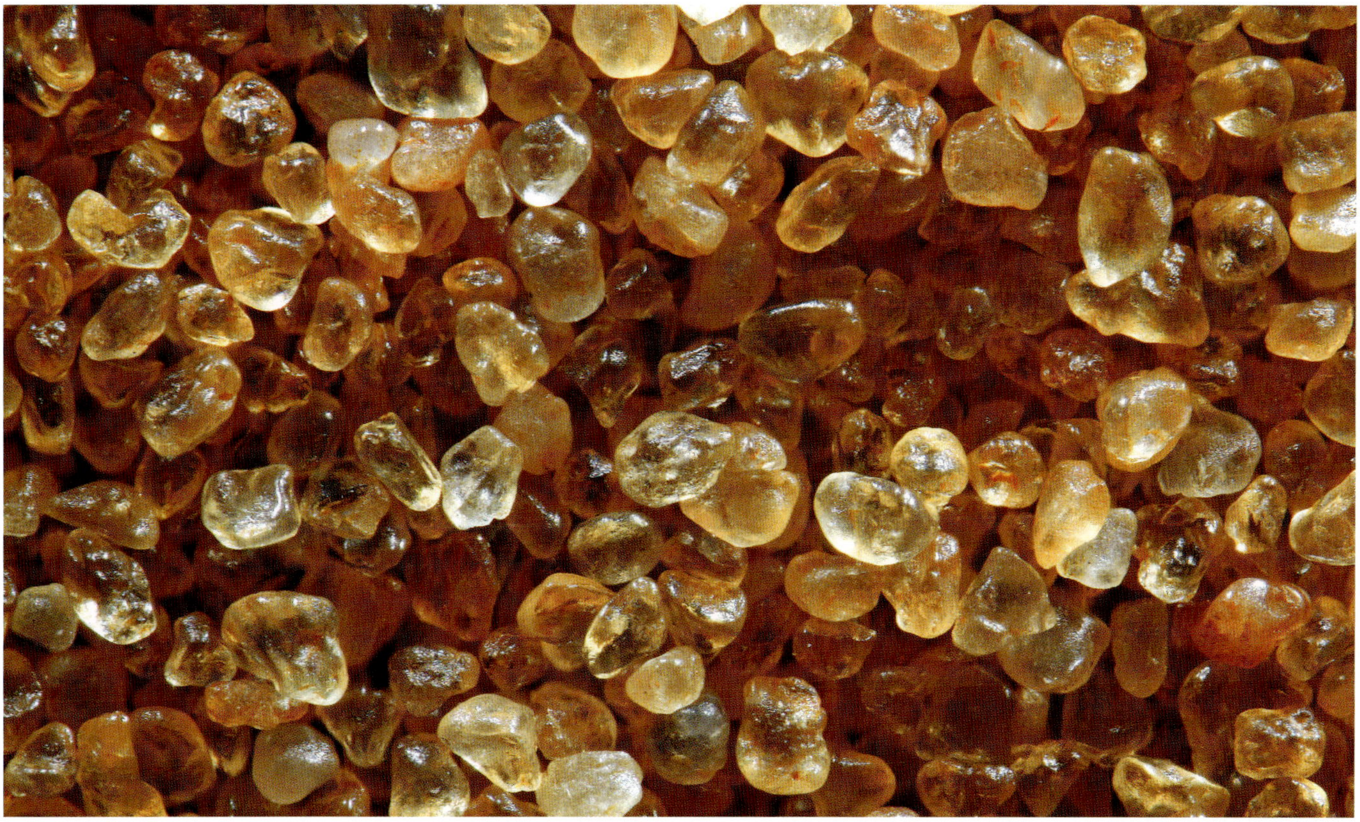

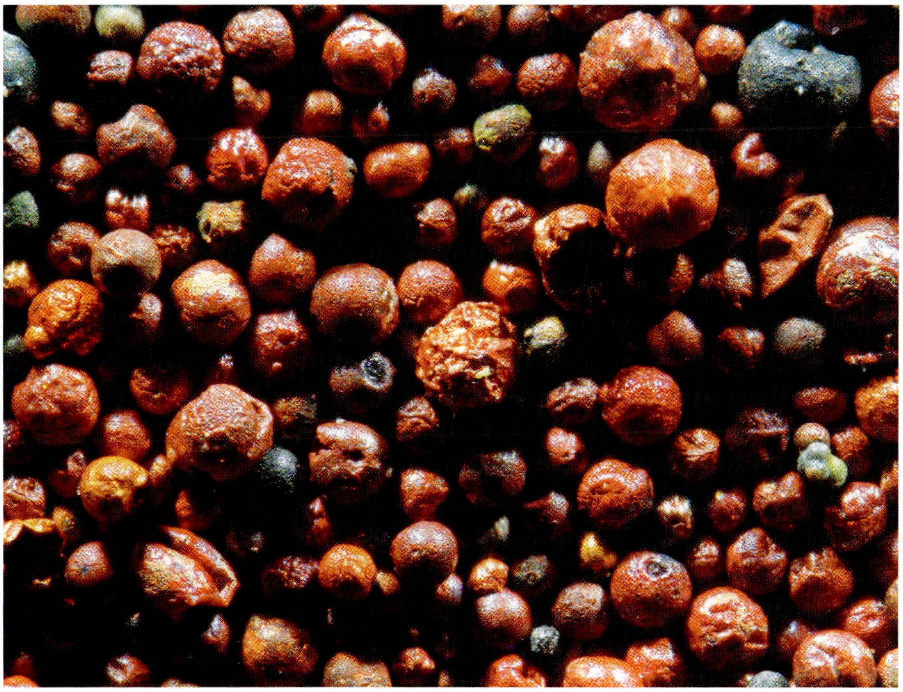

Sand from Coral Dunes, Utah, consists of rounded quartz grains that have undergone several cycles of weathering and erosion, re-formation into sandstone, and further weathering, transport, and abrasion (magnification 70x).

Grains of sand from New Mexico are made of copper that accumulated downwind of a copper smelter. The grains of copper precipitated from the smoke that belched from unregulated industrial smokestacks (magnification 50x).

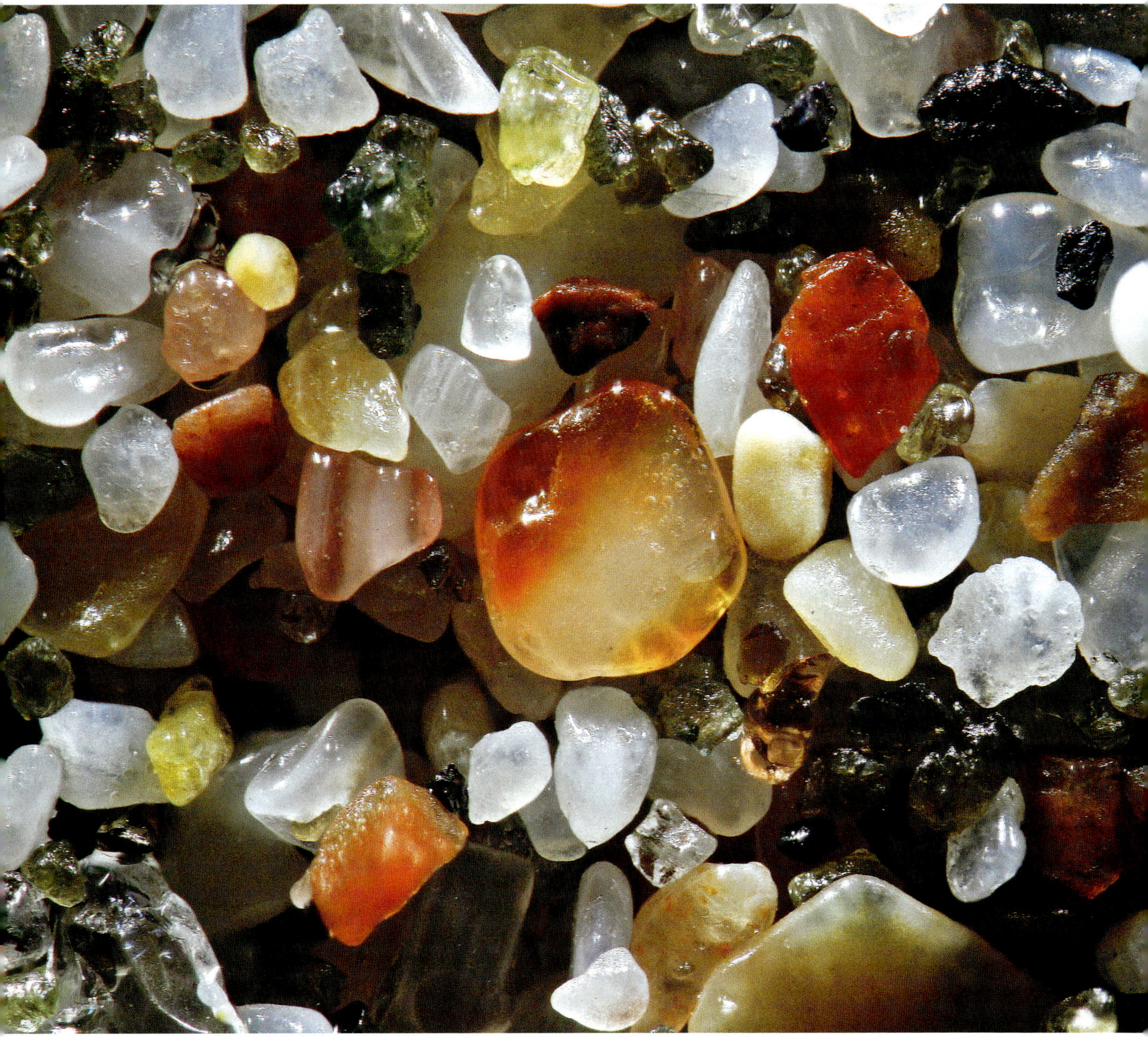

A multicolored sand sample from Flamingo Beach, Costa Rica, comprises mineral grains and shell fragments (magnification 110x).

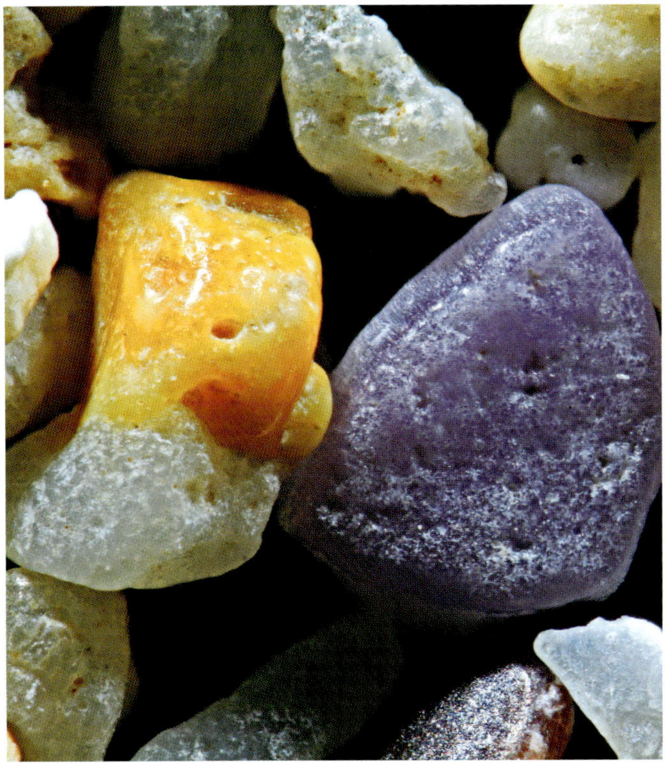

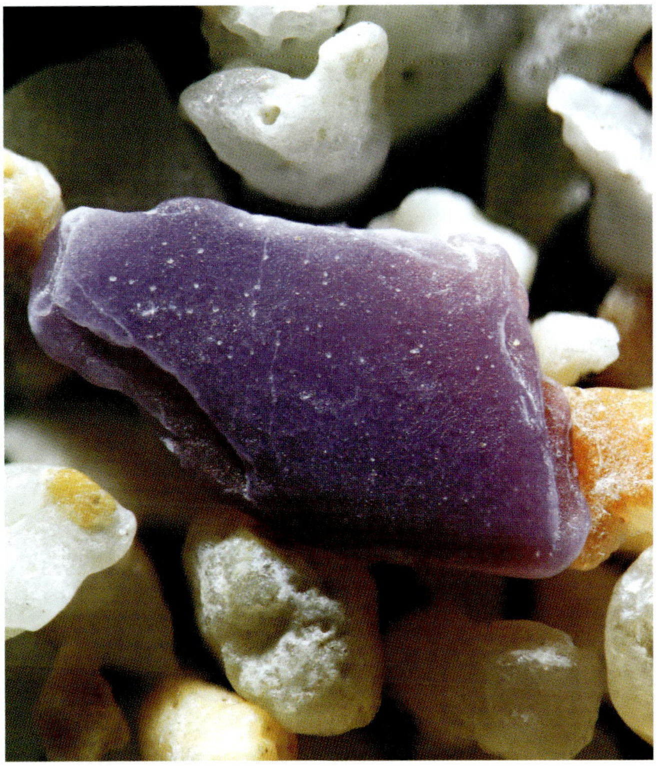

Yellow and purple sand grains were collected from Tuman Bay, Guam (magnification 140x).

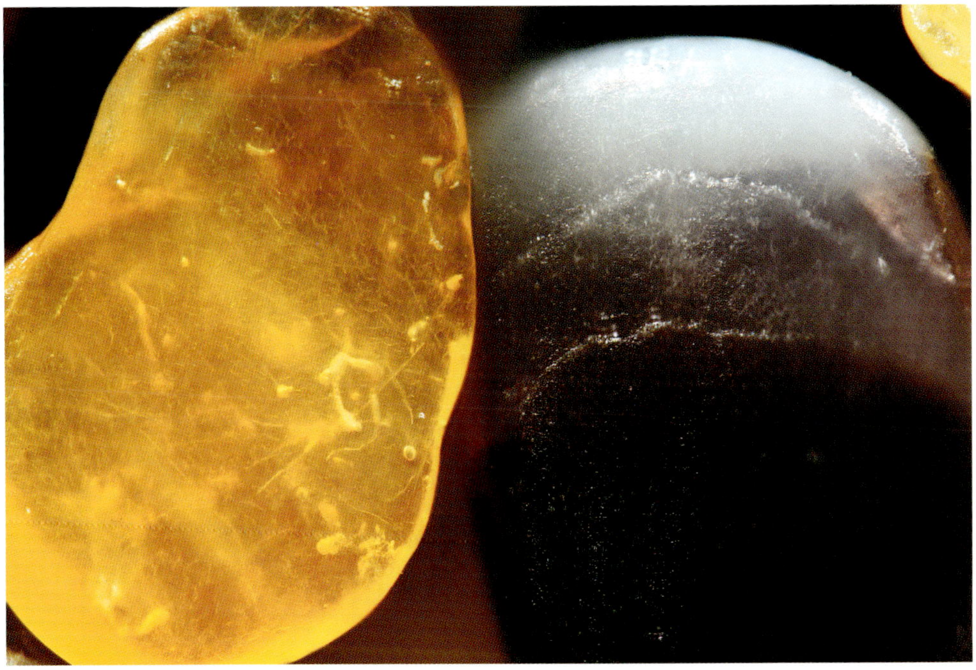

Yellow and purple sand grains from the Dingle Peninsula, Ireland. The yellow sand grain is probably citrine, a form of quartz (magnification 150x).

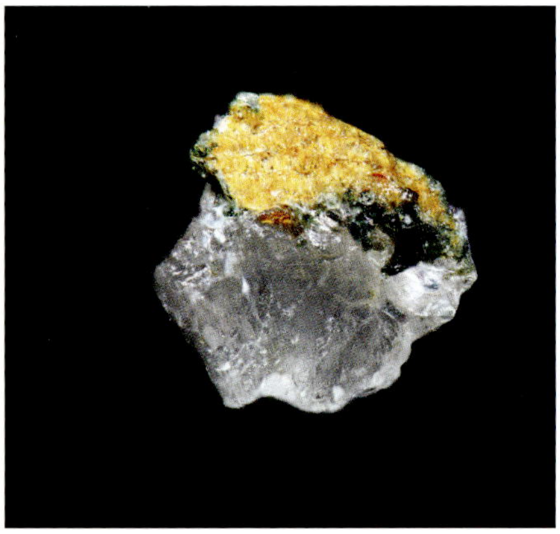

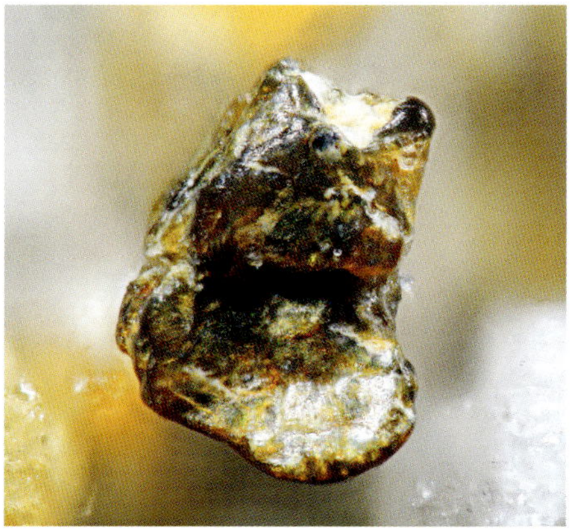

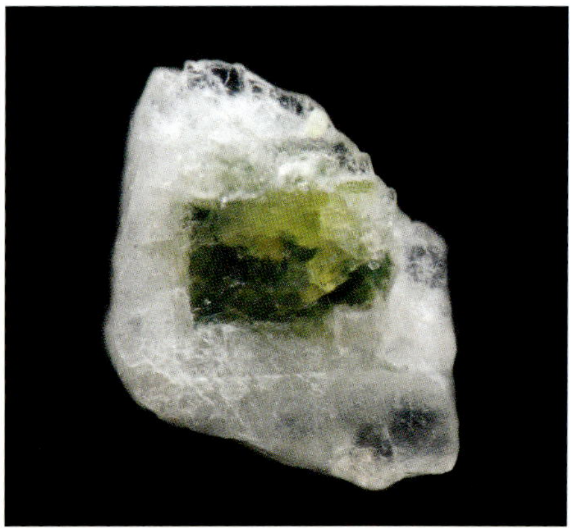

Gemlike grains of mineral sand adorn the beach at Iwate Prefecture, Honshu Island, Japan, shimmering like abstract jewels and pieces of gold. An exquisite specimen contains a rectangular green crystal embedded in the middle of a clear mineral (magnification 80x).

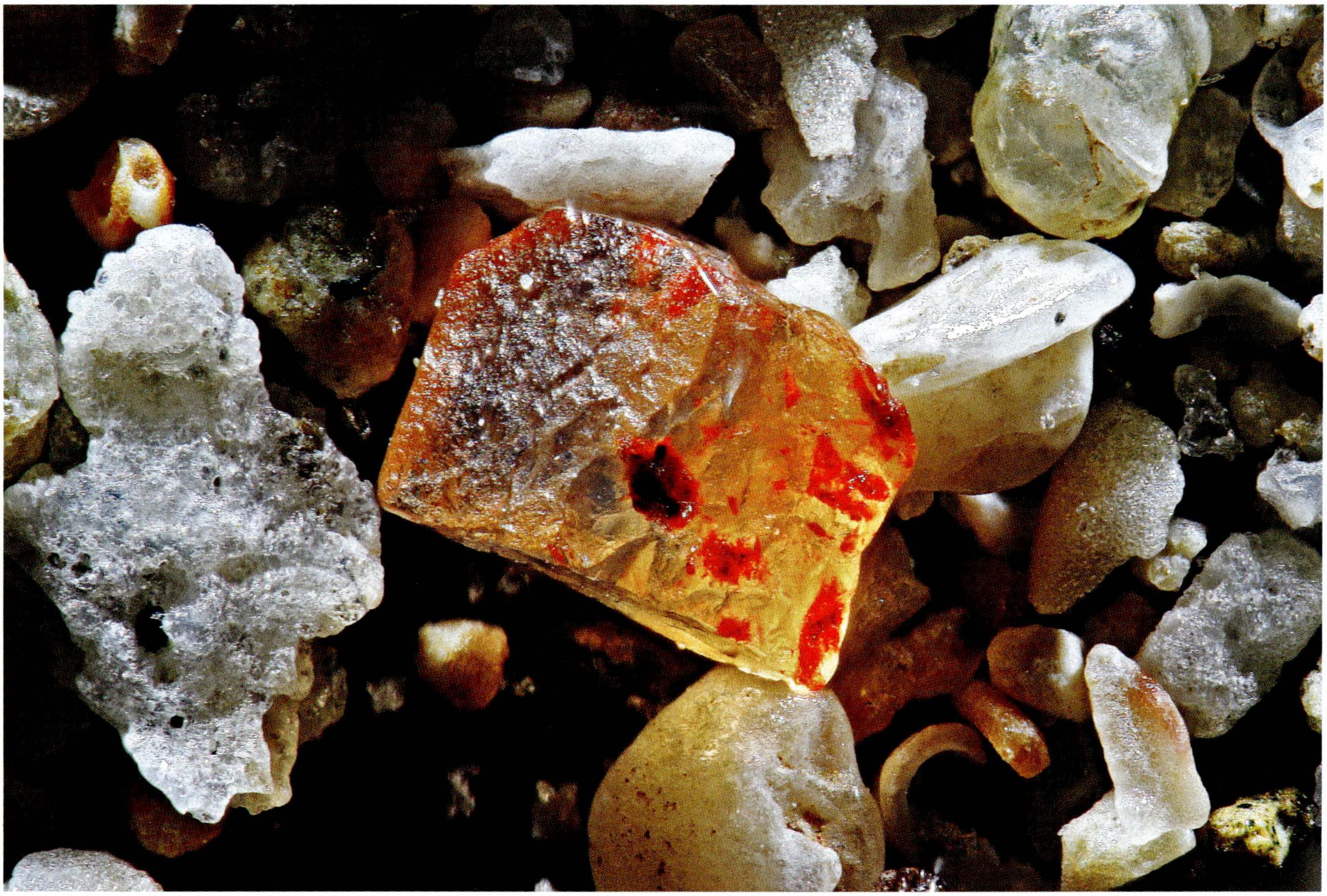

A bright sand grain from Tamarindo Beach, Guanacaste Province, Costa Rica, is made of chabazite—a glassy, cubic mineral belonging to the zeolite family of silicates (magnification 210x).

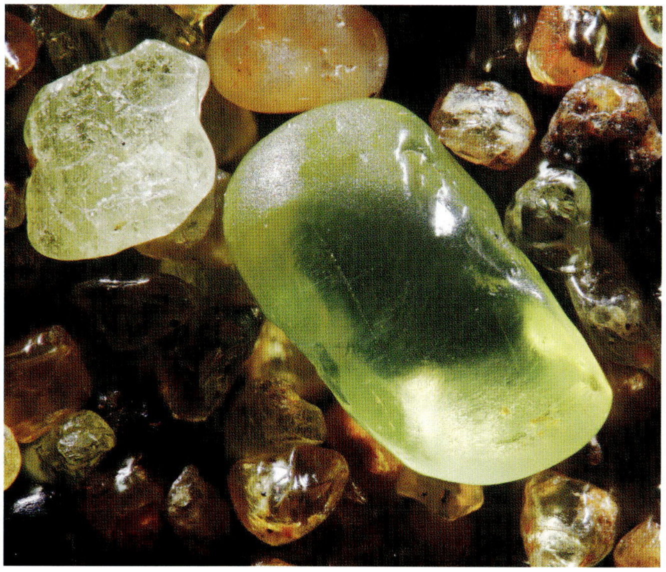

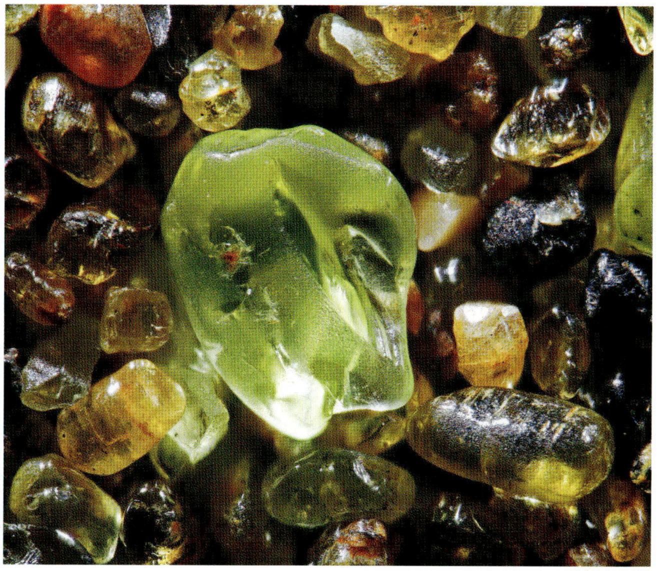

The beautiful green sand on Lumahai Beach, Kauai, Hawaii, is composed of pieces of bright-green olivine (magnification 110x).

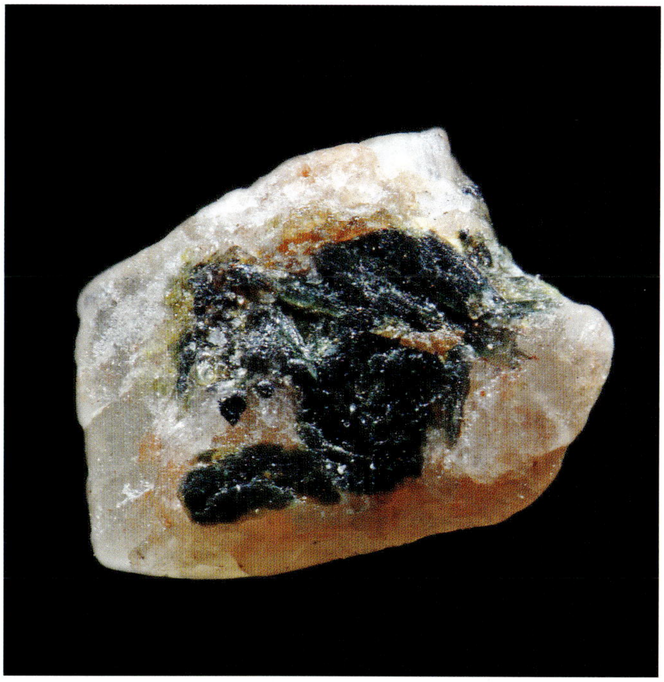

Glaciers probably transported this grain of sand from the copper-rich region in the western Upper Peninsula of Michigan to the Straits of Mackinac. The quartz in the grain gets its pink color from minute amounts of copper (magnification 125x).

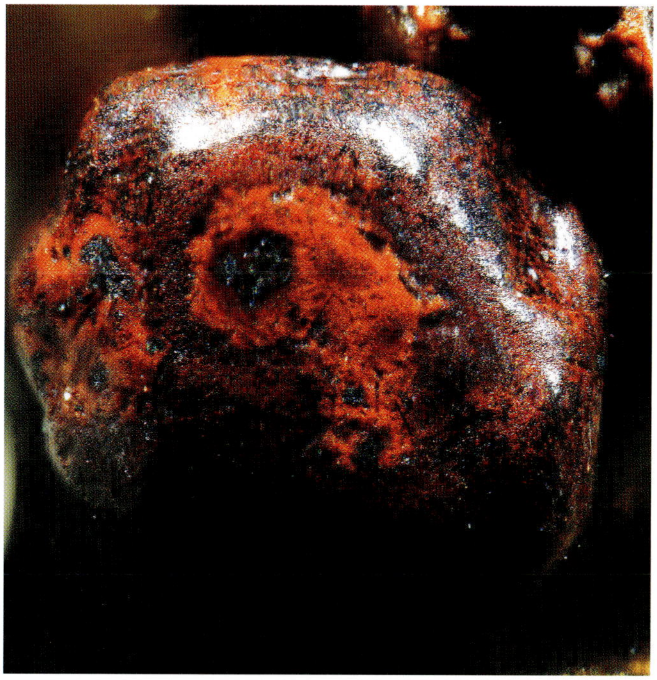

A volcanic sand grain from Makena Point, Maui, Hawaii, is decorated in red and brown (magnification 125x).

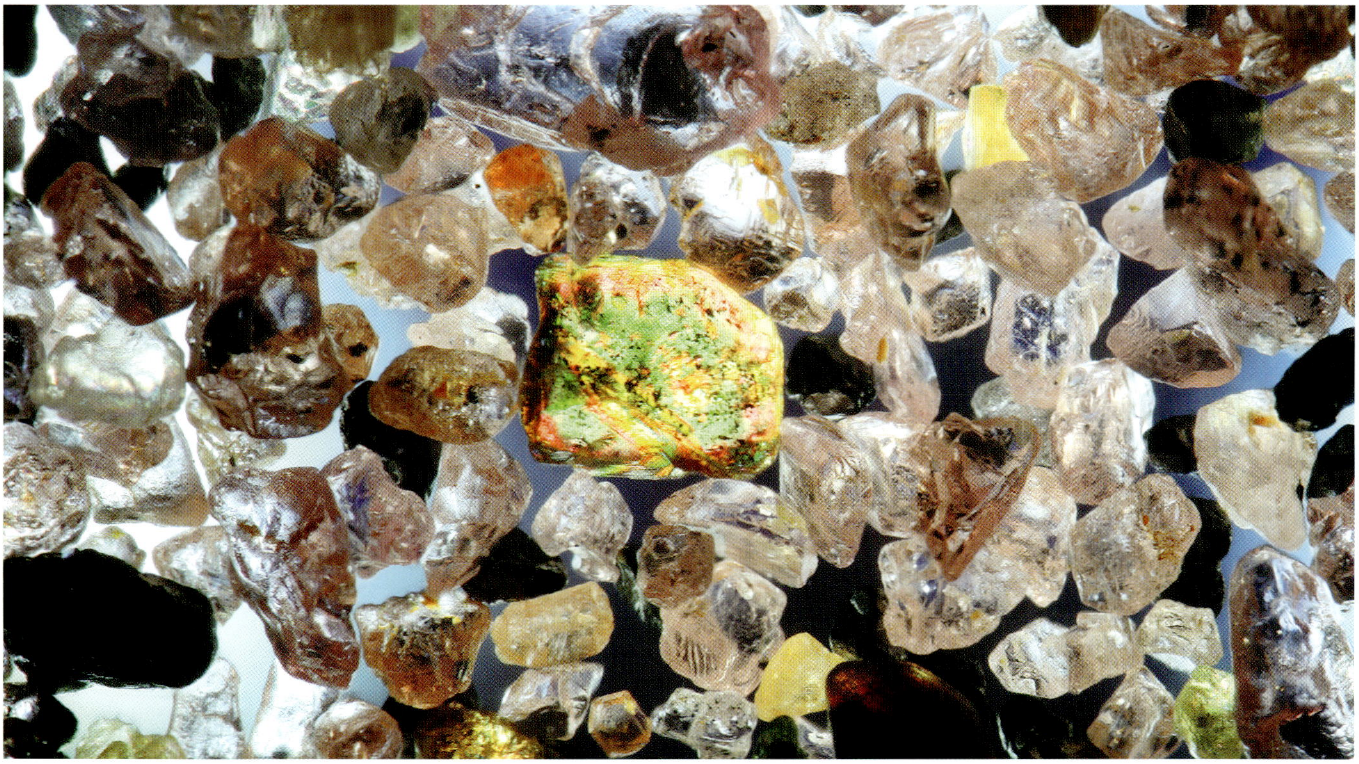

Grains of sand from Plum Island, Rowley, Massachusetts, are seen with crossed polarizing filters. Much of the sand is pink garnet, which is particularly pretty when magnified. Crossed polarizing filters create "false color" images of any sand grain that twists or rotates the plane of polarized light. The colors provide scientists with information about the crystal structure of the grain. The lighting is responsible for the rainbow effect in the central disc-shaped grain (magnification 70x).

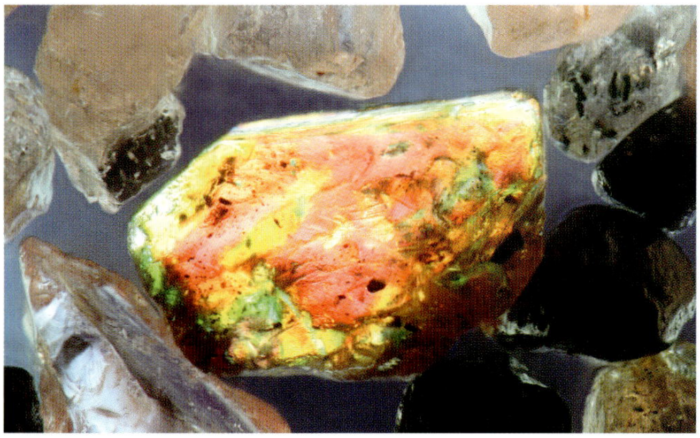

A colorful sand grain from Plum Island is surrounded by garnet and magnetite (magnification 110x).

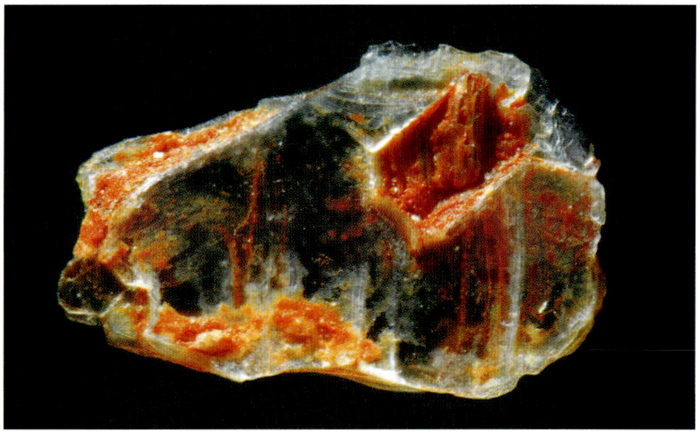

A grain of sand from Smith Mountain Lake, Virginia, appears to be mica with another mineral between the mica sheets (magnification 110x).

The Colors of Biogenic Sand

Who would have thought that a pinch of sand could contain such an assortment of color? The ocean's biological life forms take on a variety of colors. Scientists believe the patterns and colors on shells acts as camouflage against predators. Very bright colors may serve to warn other animals to stay away by broadcasting that they are poisonous or bad-tasting.

The vast array of colors found in coral reefs is due to different types of zooxanthellae algae that live symbiotically inside the coral. The algae provide the coral with nutrients and accelerate calcification. In exchange, the coral provides protection for the zooxanthellae, so they can multiply safely within the shelter of the reef. If the coral is stressed, the algae may depart along with the color, leaving the coral white and bleached-looking. The bleaching of coral occurs in response to environmental stresses such as overexploitation, nutrient overload from agricultural runoff, increased ultraviolet radiation from ozone depletion of the atmosphere, increased alkalinity of the ocean, and increased ocean temperature from global warming. The reef then typically dies without the colorful algae.

Reports indicate that 20 percent of the coral reefs on Earth have been destroyed over the past few decades and another 50 percent are in poor health. The decline of coral reefs will have dramatic effects on the fishing industry and tourism. Coral reefs are the rainforests of the ocean, and it is essential that we protect and nourish them.

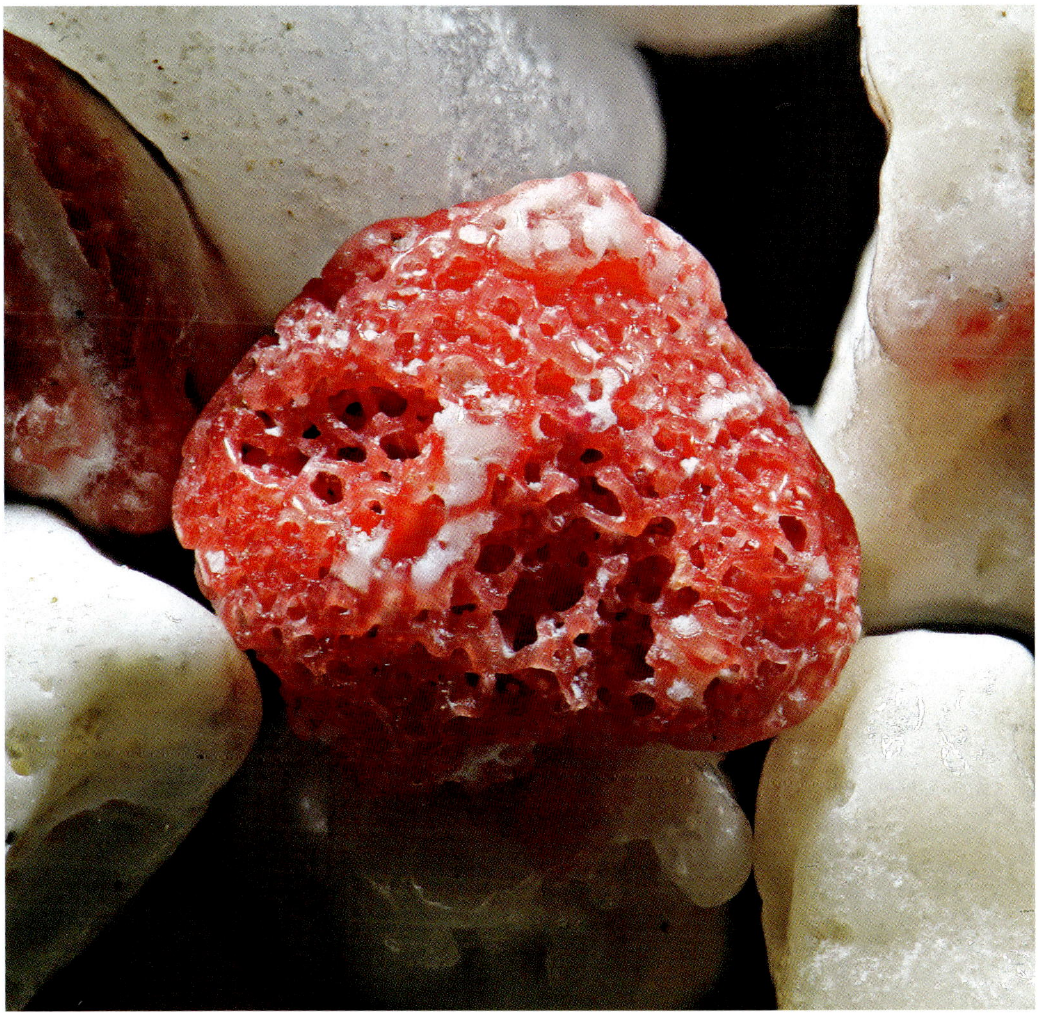

A grain of sand is made of pink coral that lives in the warm tropical waters around French Bay, Bahamas (magnification 190x).

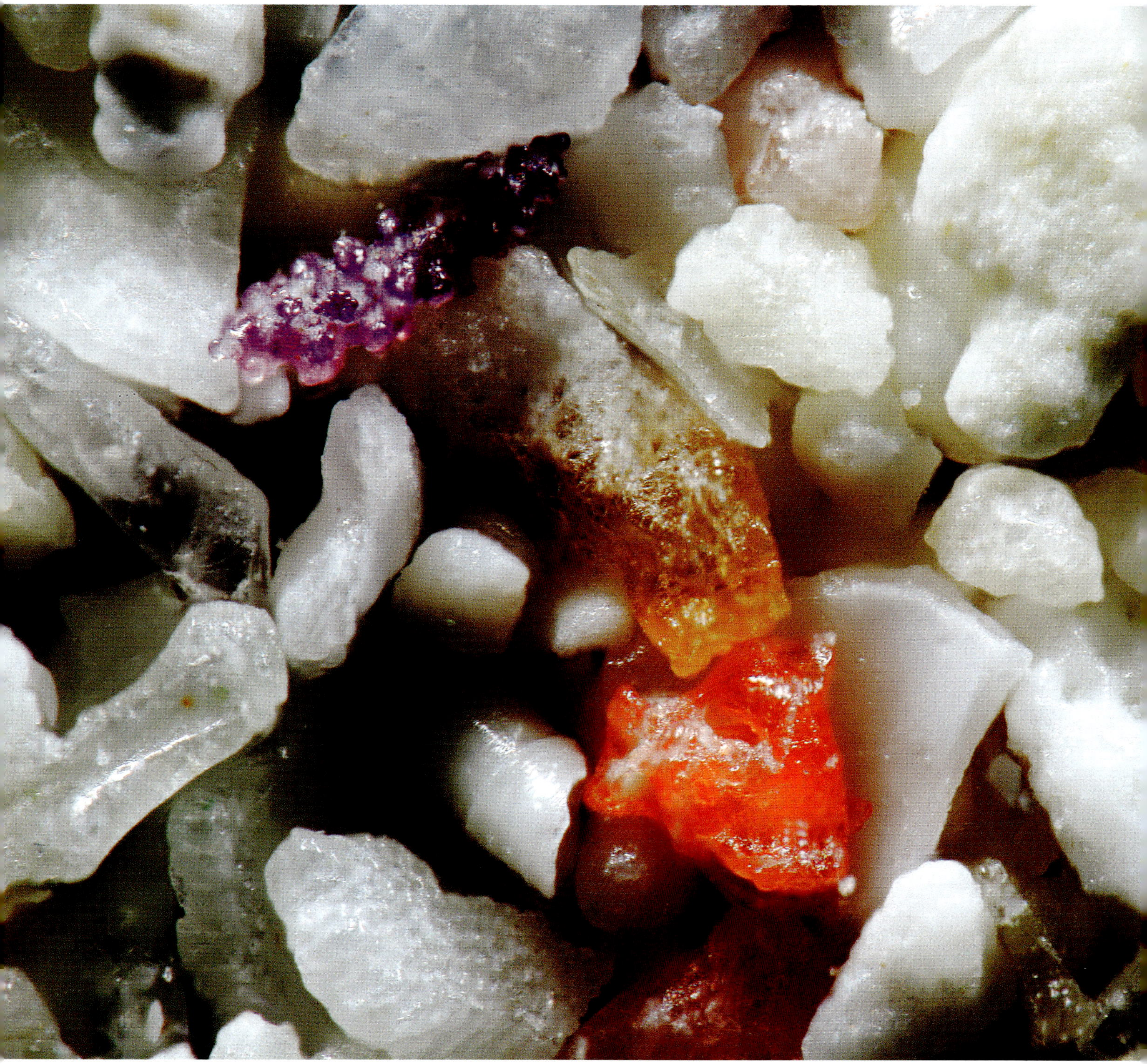

Look closely at these colorful bits of sand from Southampton, Bermuda, and you'll see fragments of tropical coral, shells, and forams (magnification 260x).

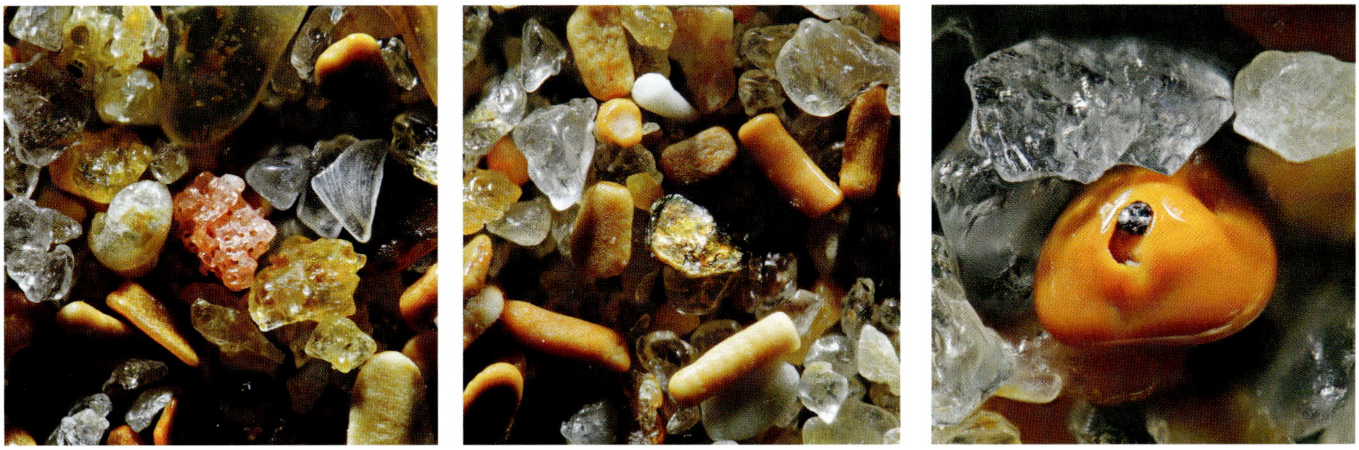

Images of sand from Geriba Beach, Brazil, reveal (*left*) shell fragments, pink coral, and clear quartz grains (magnification 35x); (*middle*) tan-colored coral, clear quartz, and a flake of mica in the center (magnification 35x); and (*right*) quartz sand surrounding a beautiful orange shell fragment (magnification 80x).

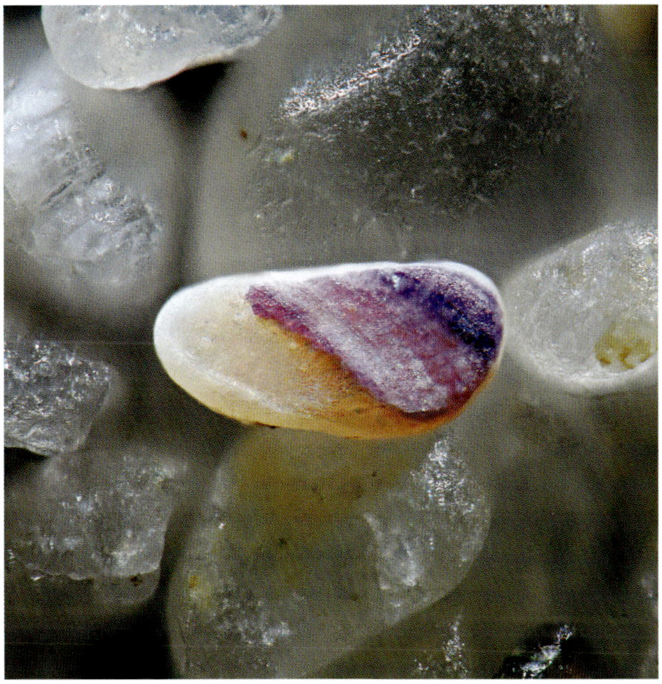

Sand from Glen Beach, Cape Town, South Africa, shows a beautifully eroded shell fragment surrounded by rounded quartz grains (magnification 125x).

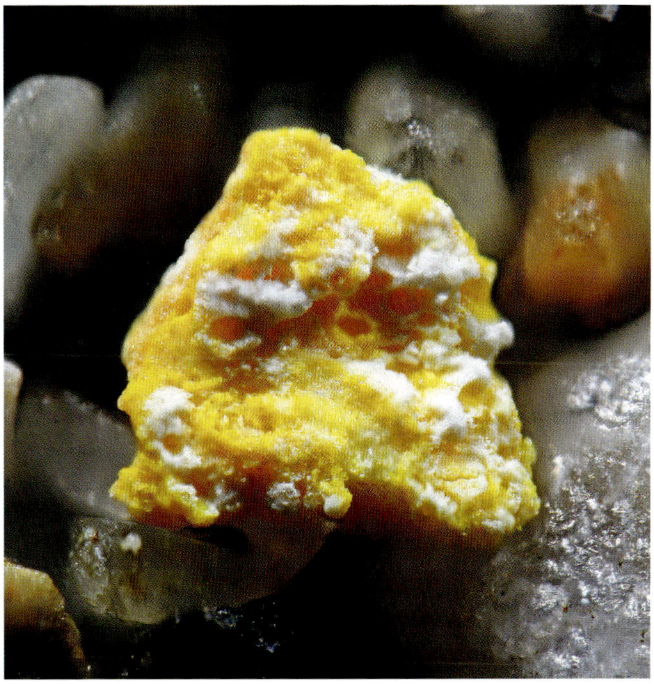

Sand from Marina di Castagneto Carducci, Italy, includes a bright chip of yellow coral in the process of being dissolved and eroded (magnification 125x).

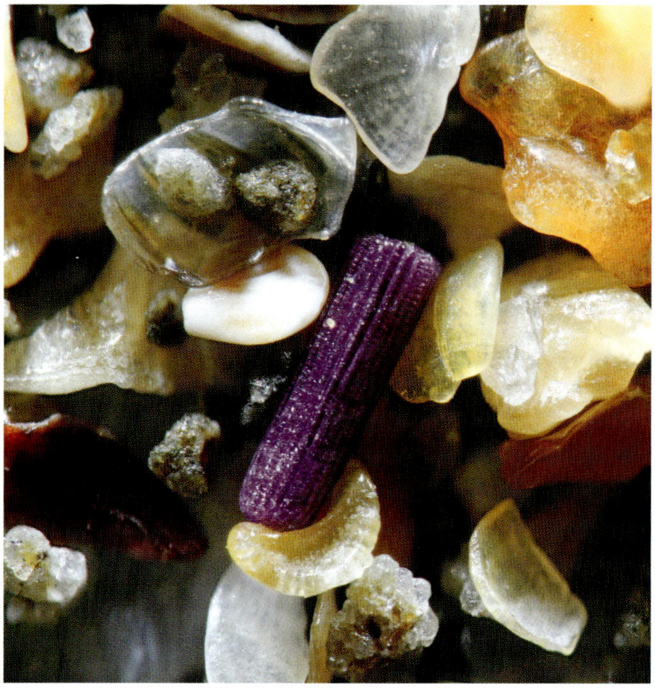

Sand from Fanore, Ireland, contains a bright-purple fragment of sea urchin spine along with various shell fragments (magnification 125x).

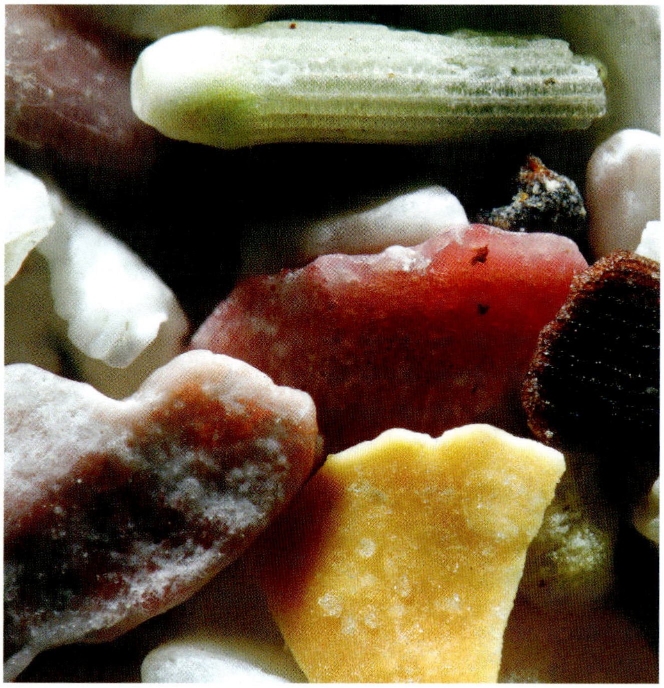

Sand from the Galápagos Islands contains two beautifully colored pieces of sea urchin spine, one at the top of the frame and one at the right (magnification 125x).

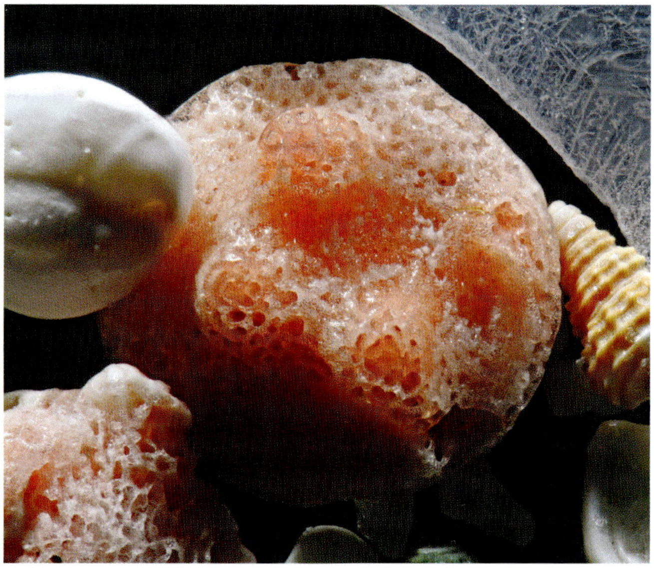

Sand from the island of Corsica, France, in the Mediterranean features a piece of pale-pink coral, a micro shell, shell fragments, and a small foram (magnification 95x).

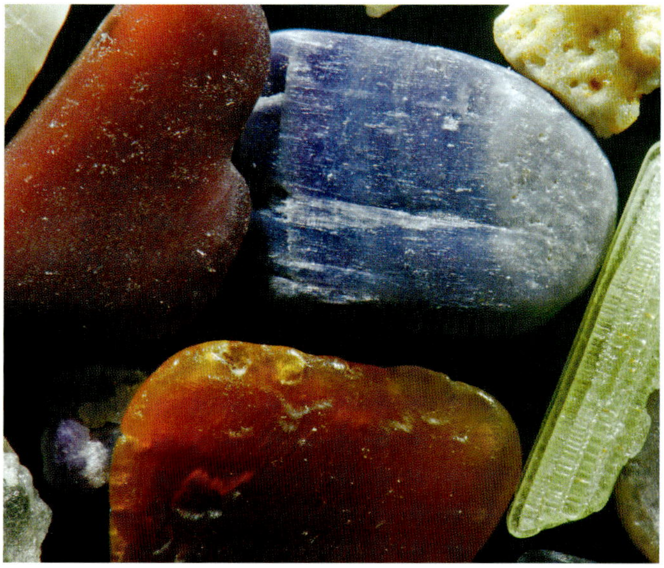

Mineral grains and a light-green piece of sea urchin spine surround a piece of blue kyanite in sand from the Isle of Shoals, Maine (magnification 95x).

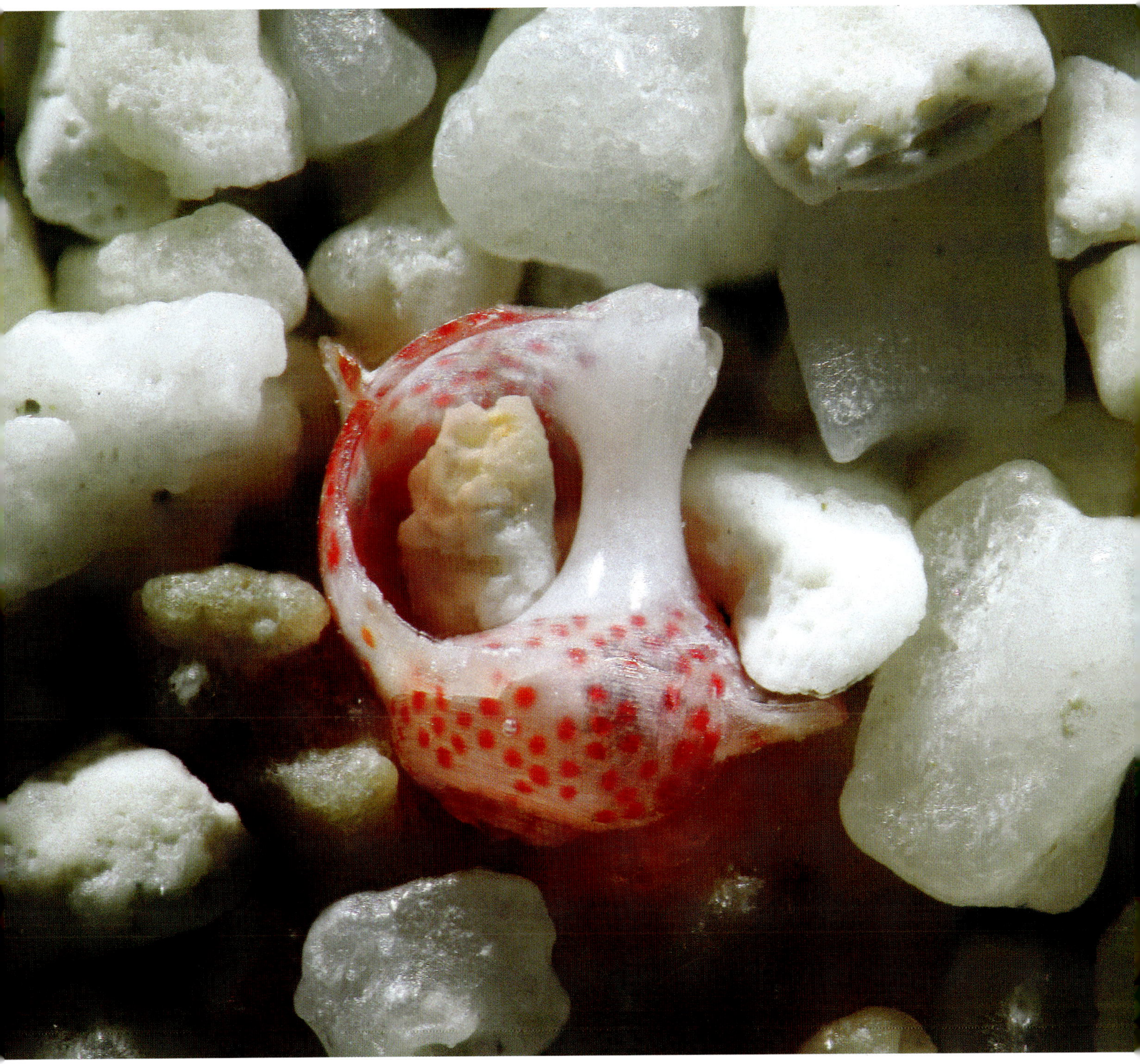

Tropical beach sand from Utila, Honduras, contains a beautifully decorated microscopic shell fragment (magnification 260x).

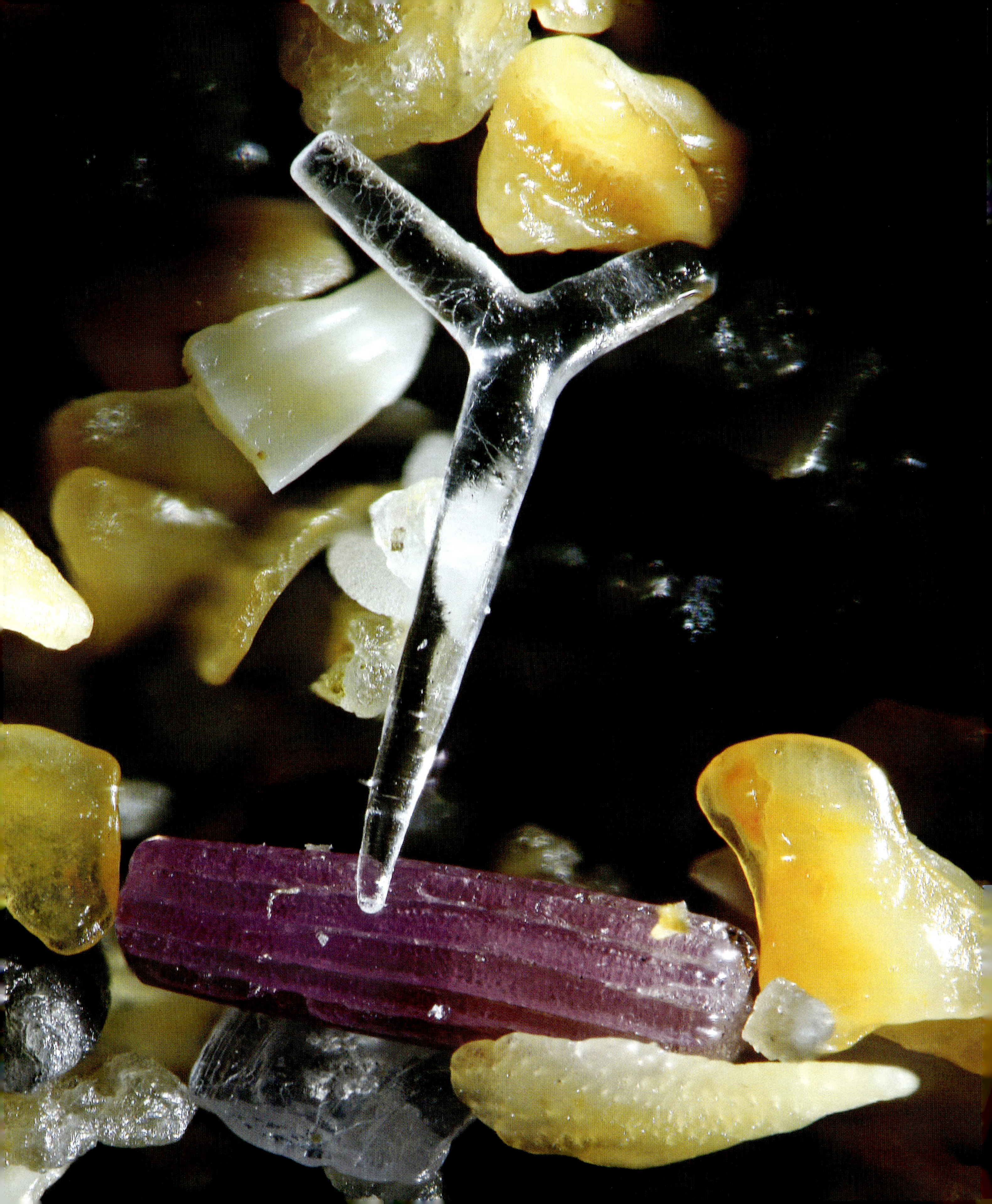

CHAPTER 5

Shapes of Sand Grains

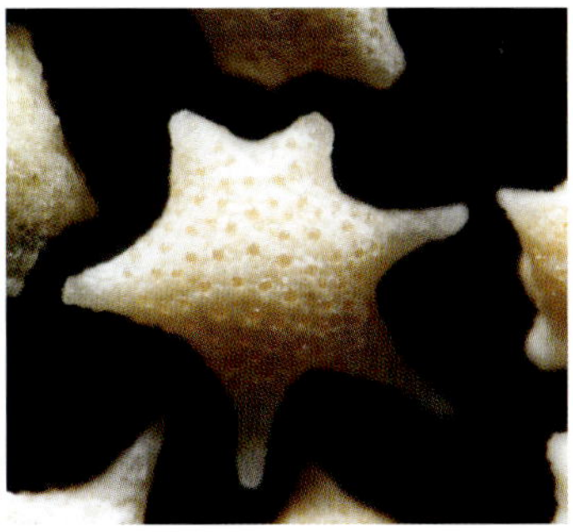
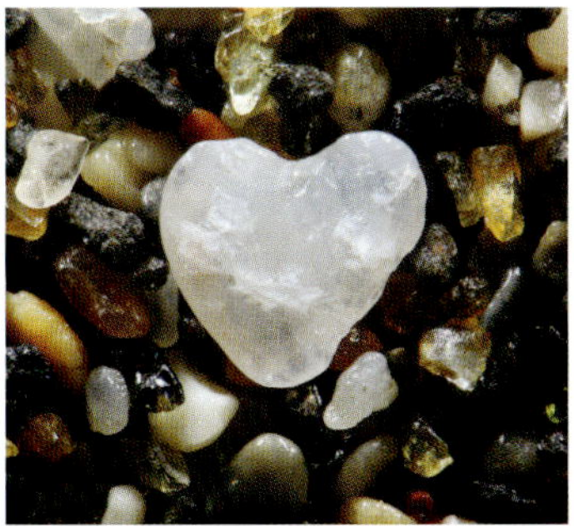
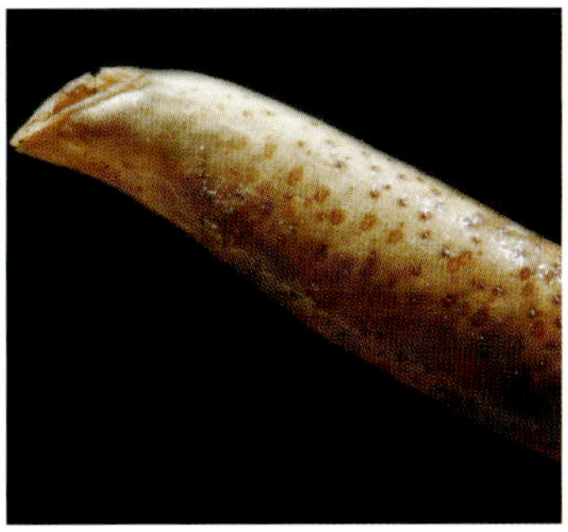

Individual sand grains come in a wide variety of shapes, each more remarkable than the next. The images in this chapter reveal the wonderful shapes individual grains can take, such as that of the striking sponge spicule. The amazing thing is that no two grains of sand have exactly the same shape. Each is in the process of changing and eroding from the moment it becomes an individual grain of sand. Every grain has a history throughout time and space, and much of that history can be read through the microscope.

The shapes of the mineral components of sand are particularly beautiful. As magma cools and solidifies into rock, it often forms magnificently shaped microcrystals. Each crystal begins to form from a nucleation site and grows into a three-dimensional shape with numerous facets. You can see interlocking crystal shapes through the microscope where individual microcrystals have merged into intricate sculptural forms.

Sand from Fanore, Ireland, contains a beautiful three-pronged sponge spicule and an intricate bit of purple sea urchin spine. The 500-meter-deep coral reefs around Ireland play host to a rich ecosystem and are home to many marine species, including sea urchins, clams, sea stars, sponges, and crabs. (magnification 300x).

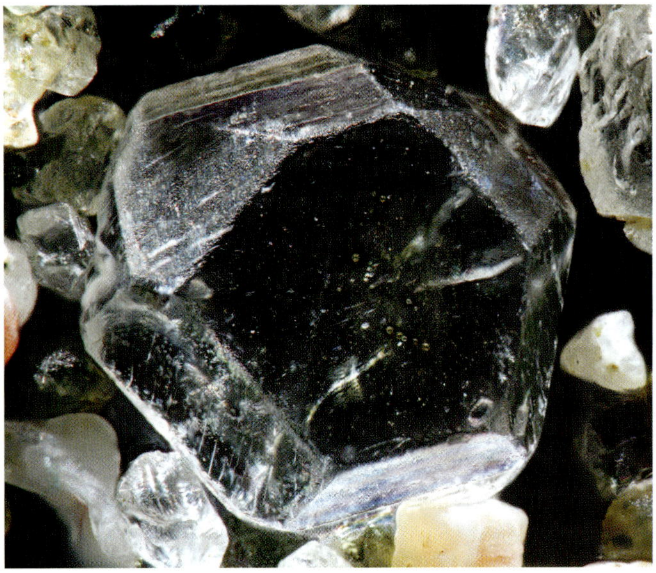

Sand from Zushi Beach, Japan, contains what looks like a sapphire crystal. The crystal is larger than the surrounding grains and has survived eroding because of its hardness and quality (magnification 105x).

The roundness of sand grains reflects the degree of erosion the grain has experienced. The rounder the grain, the more cycles of erosion, transport, deposition, and cementation the grain has gone through. Sand grains made of quartz, a hard mineral resistant to chemical and mechanical erosion, may have gone through these cycles many times.

How uniform in size are the grains of sand on a particular beach? The size of sand grains is related to how far the sand has moved from its source of origin, as well the way the sand was transported. The sorting of sand grains into different sizes depends on the carrying power of the transport medium and the distance the sand is moved. In general, sand grains transported a short distance are less sorted than grains that have traveled a great distance.

Landslides and glaciers are powerful mediums that transport materials relatively short distances, and these mediums typically deposit poorly sorted sediments. In other words, glaciers can deposit boulders, pebbles, and sand all in the same short distance. Rivers and streams carry sediments longer distances, and they tend to sort out the grains by size. High-powered waves or rushing rivers can move large particles along with the small particles, whereas slow-moving water carries only fine particles.

Wind produces well-sorted sand grains. The velocity of the wind is critical in determining the size of the grains it can carry. Dust moves effectively at even low wind speeds. Larger sand particles bounce along when transported by heavy winds. Strong winds usually deposit dunes of sand composed of uniformly sized grains.

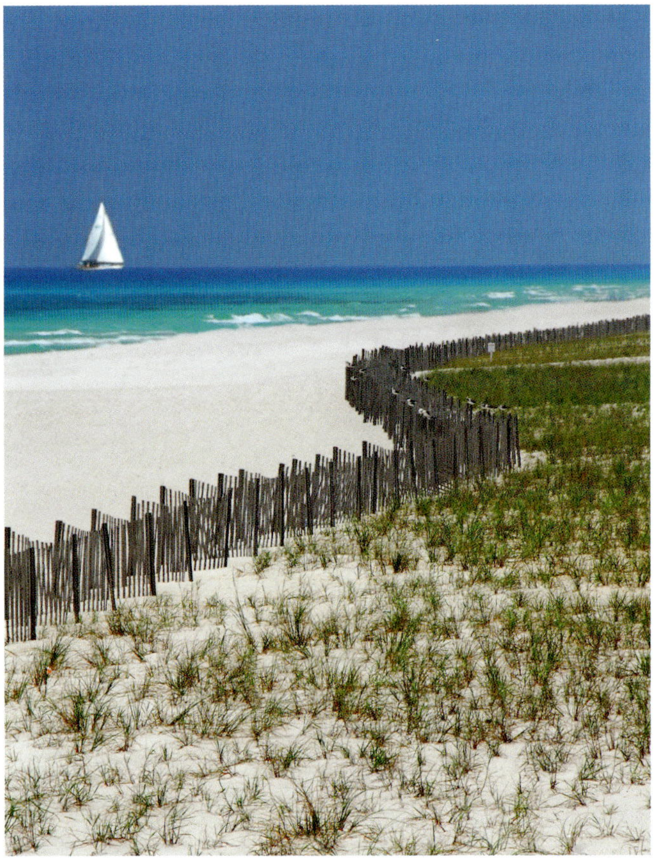

A weather-beaten fence and shore grasses capture windblown sand grains, creating healthy dunes on Pensacola Bay, Florida. *Cheryl Casey, Shutterstock*

The Shapes of Mineral Sand

The shapes of mineral grains often aid in identifying the origins of sand on a beach. Grains showing great uniformity in roundness have experienced multiple cycles of erosion and transport. The sands on the beaches of Copacabana, Brazil, Leisure Bay, South Africa, and Seal Bay, Australia, contain rounded and polished quartz grains—a testament to their origins as continental rock with a long history of erosion and transport.

Certain minerals are easily recognizable. A magnet will attract grains of magnetite and other iron-rich minerals. Mica flakes into layers and these stand out even without magnification; under the microscope, you can easily see the flakelike structure of mica in the sand grains. Garnets in sand often occur as rounded grains, having eroded from their original dodecahedral (twelve-sided) crystals.

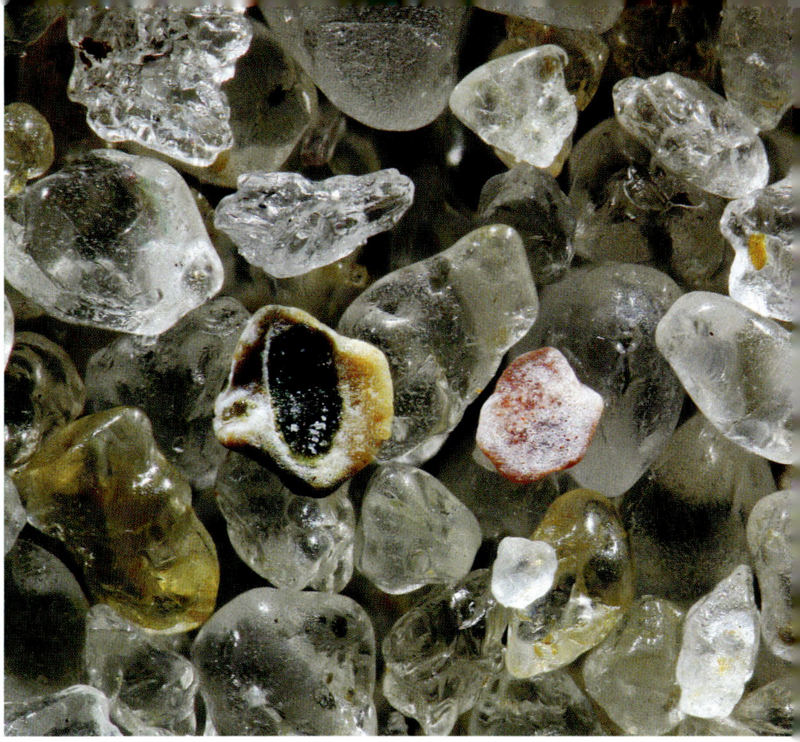

Sand from Copacabana Beach, Brazil, consists of a dark shell fragment amid rounded and polished clear quartz. The purity and roundedness of the grains stand as a record of years of erosion and transport (magnification 60x).

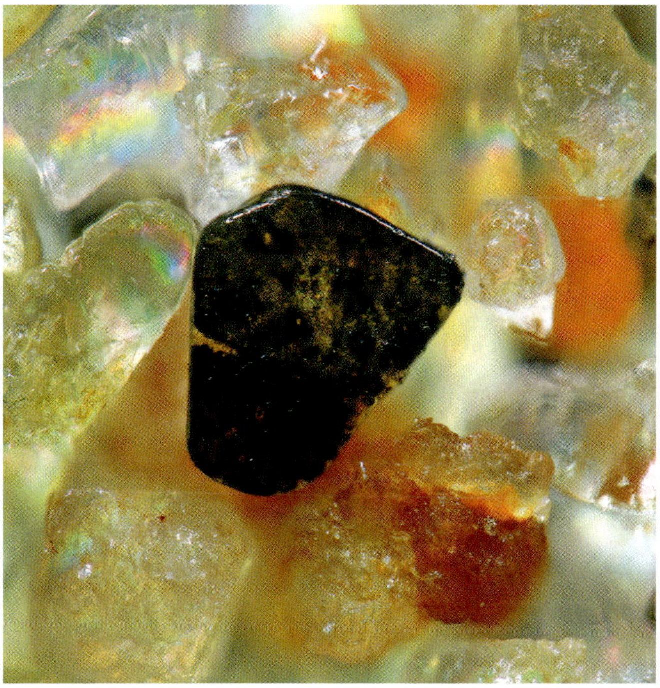

Clear quartz sand grains from Leisure Bay, South Africa, on the Indian Ocean take on a rainbow appearance when viewed using crossed polarizing filters (magnification 125x).

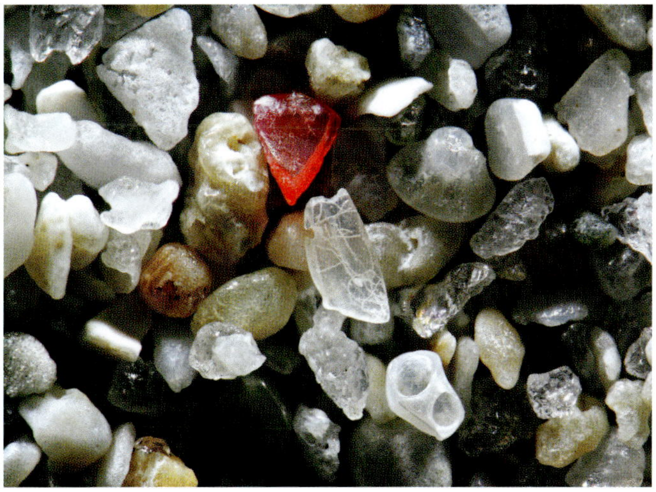

Some interestingly shaped grains can be seen in sand from Seal Bay, Australia. The quartz and feldspar sand grains originated as continental rock. Many of the grains are rounded, indicating that they have endured numerous years of erosion and transport. The red grain may be a tiny ruby (magnification 45x).

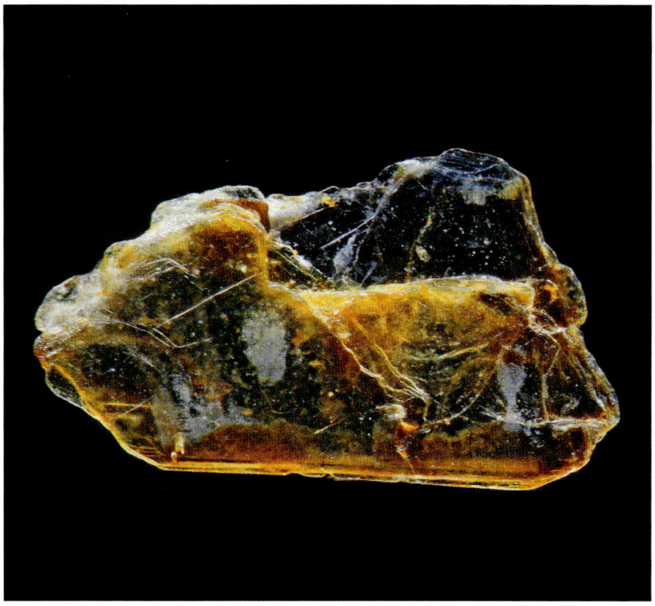

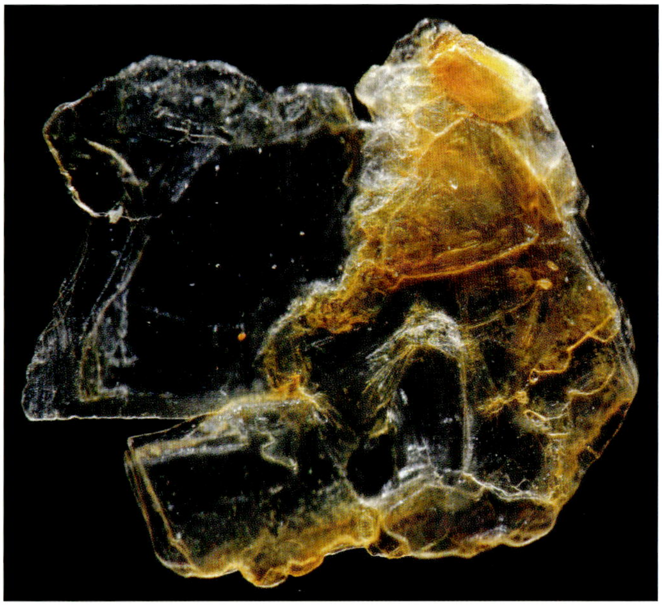

Sand from Smith Mountain Lake, Virginia, contains mica, a common mineral in igneous and metamorphic rocks. Mica is a soft mineral with a hardness of 2 to 2.5 but a density of 2.8, heavier than quartz. In spite of its softness, it is resistant to water erosion. In water, mica rests flat or floats but doesn't roll, thus resisting rounding or smoothing. Rolling is the mechanism that rounds and smoothes most sand grains (*left*, magnification 45x; *right*, magnification 100x).

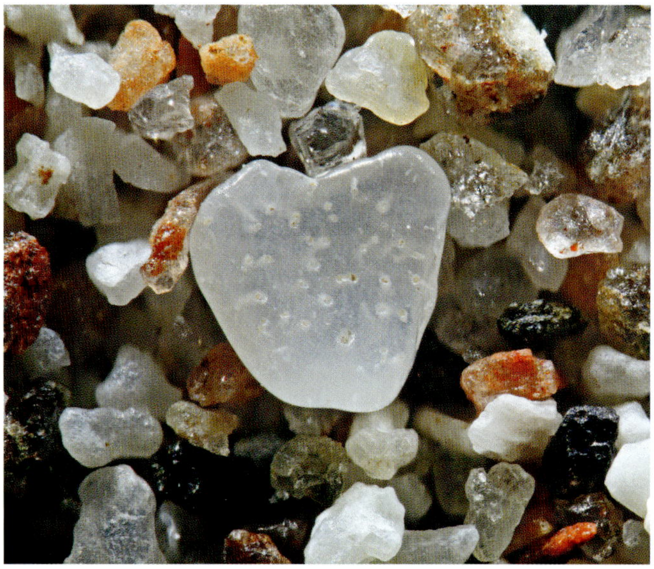

An apple-shaped rounded fragment stands out among small angular grains of sand from Bahia de Los Angeles, Baja California, Mexico (magnification 45x).

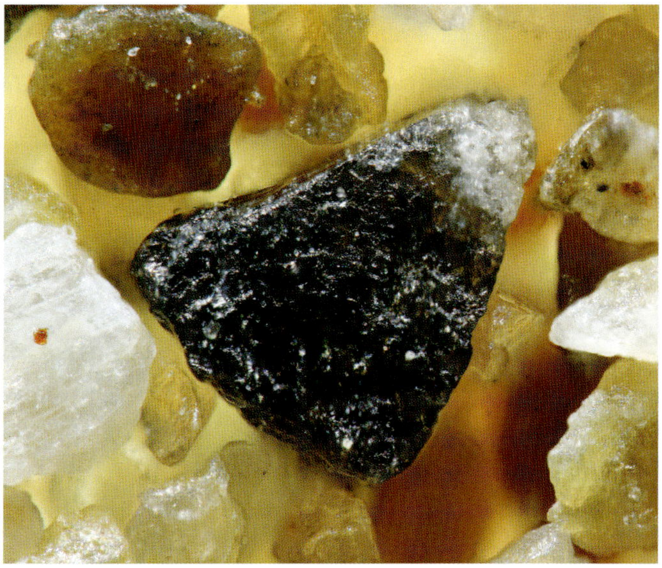

Sand from Barcelona, Spain, contains both angular and rounded grains, including a pie-shaped grain of sand (magnification 105x).

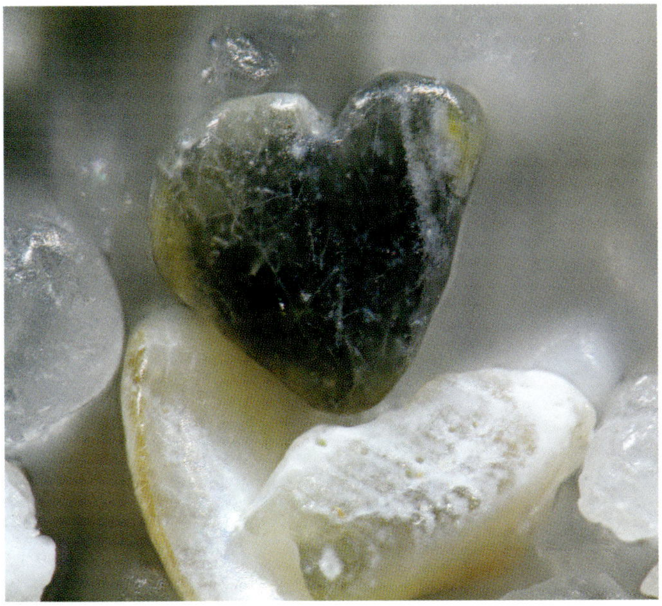

Sand from Glen Beach, Cape Town, South Africa, shows a heart-shaped grain amid polished and rounded grains. The grains weathered from rocks that originated on the African continent (magnification 110x).

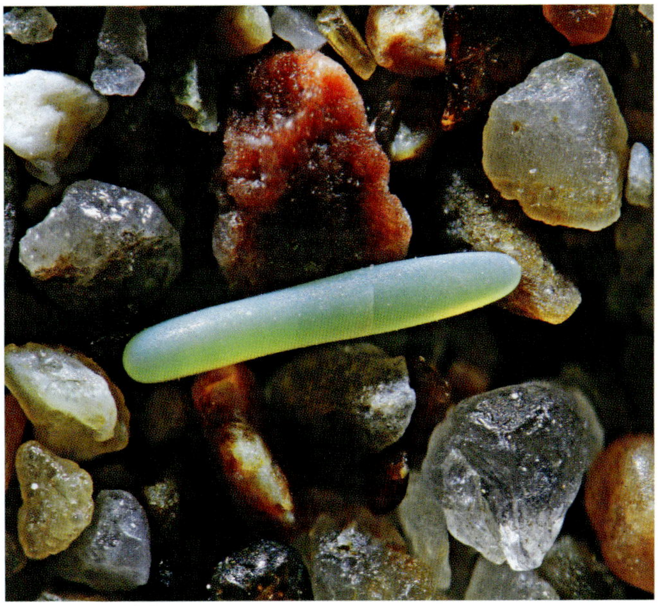

An elegant, light-green sand grain from Marina di Castagneto Carducci, Italy, has an unusual shape and color (magnification 105x).

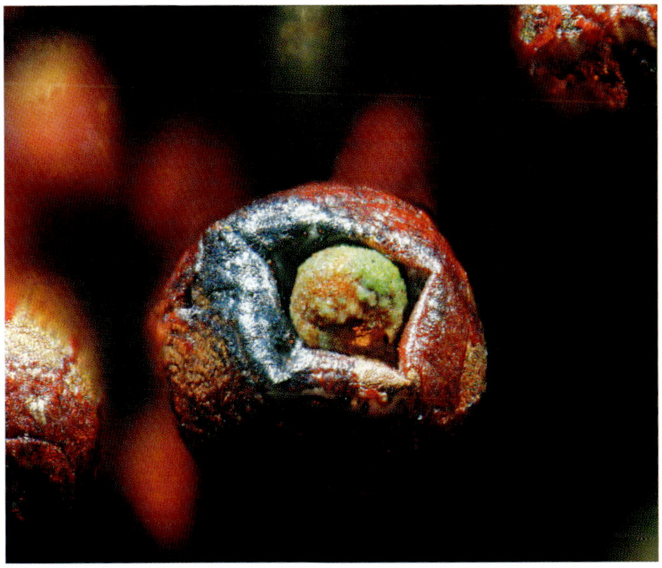

A small grain of copper sand is impacted into a larger copper grain. The copper sands precipitated from an unregulated smelter in New Mexico (magnification 110x).

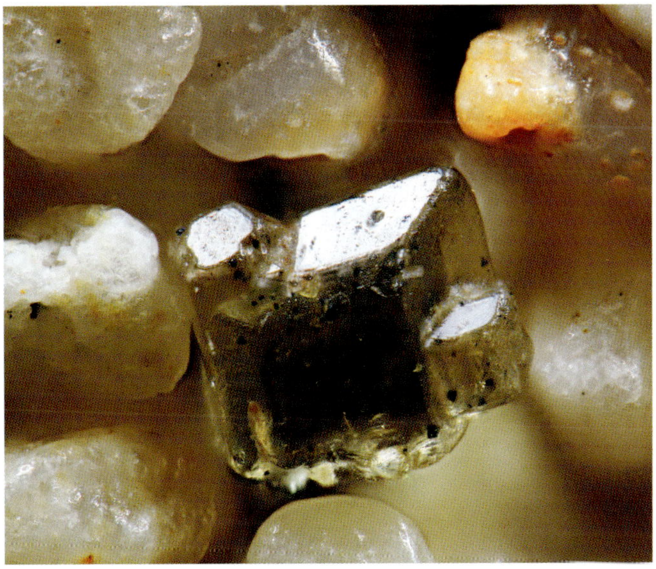

A chabazite crystal was found in sand from Diamond Head, Oahu, Hawaii (magnification 110x).

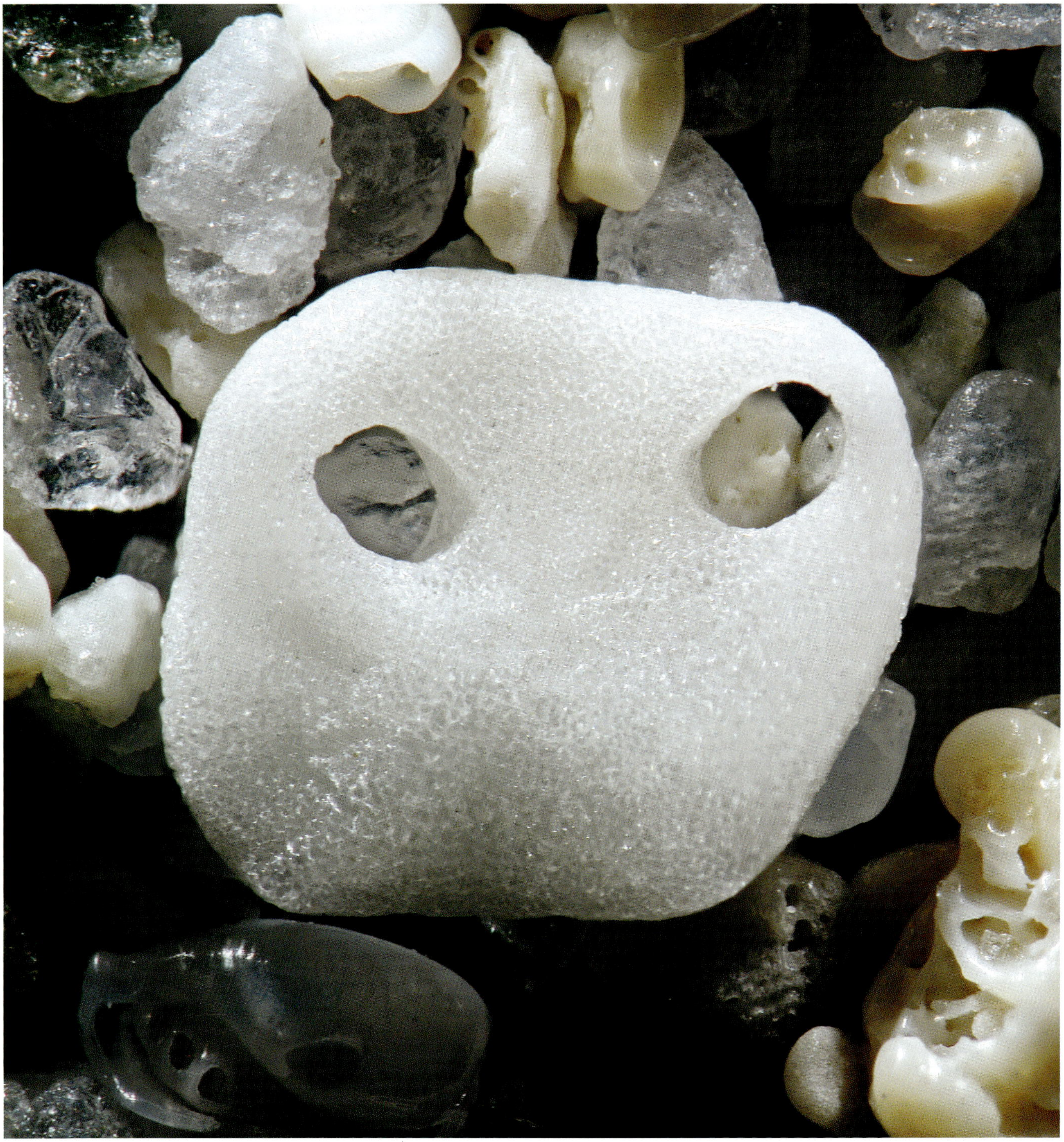

A grain of sand from the island of Corsica, France, looks like a mask (magnification 210x).

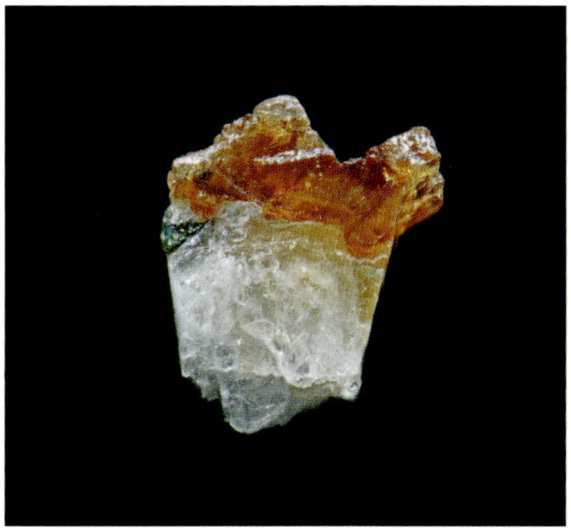
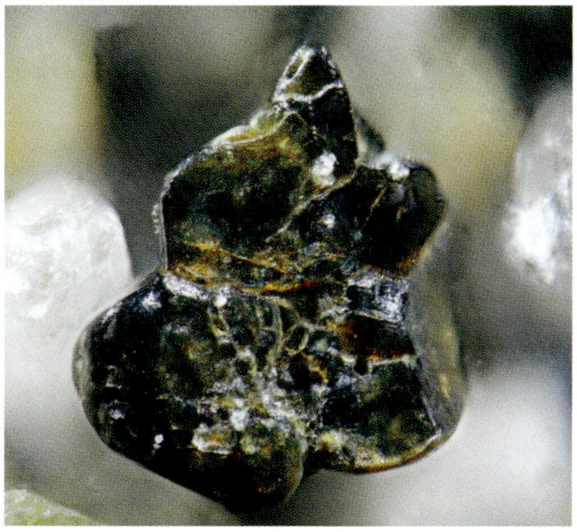
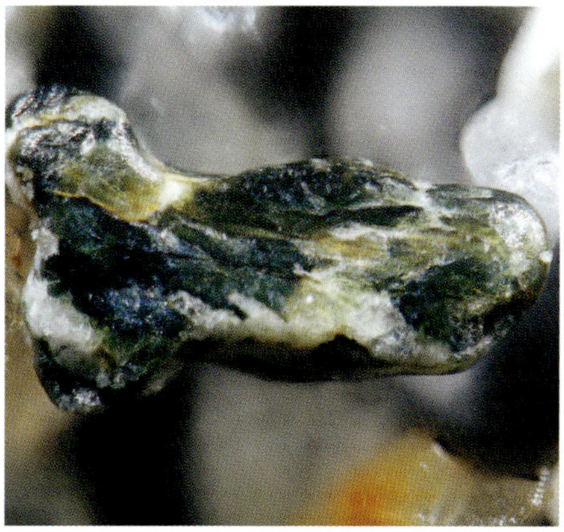

Beautiful mineral grains can be seen in sand from Iwate Prefecture, Honshu Island, Japan, a region with abundant mineral resources (magnification 65x).

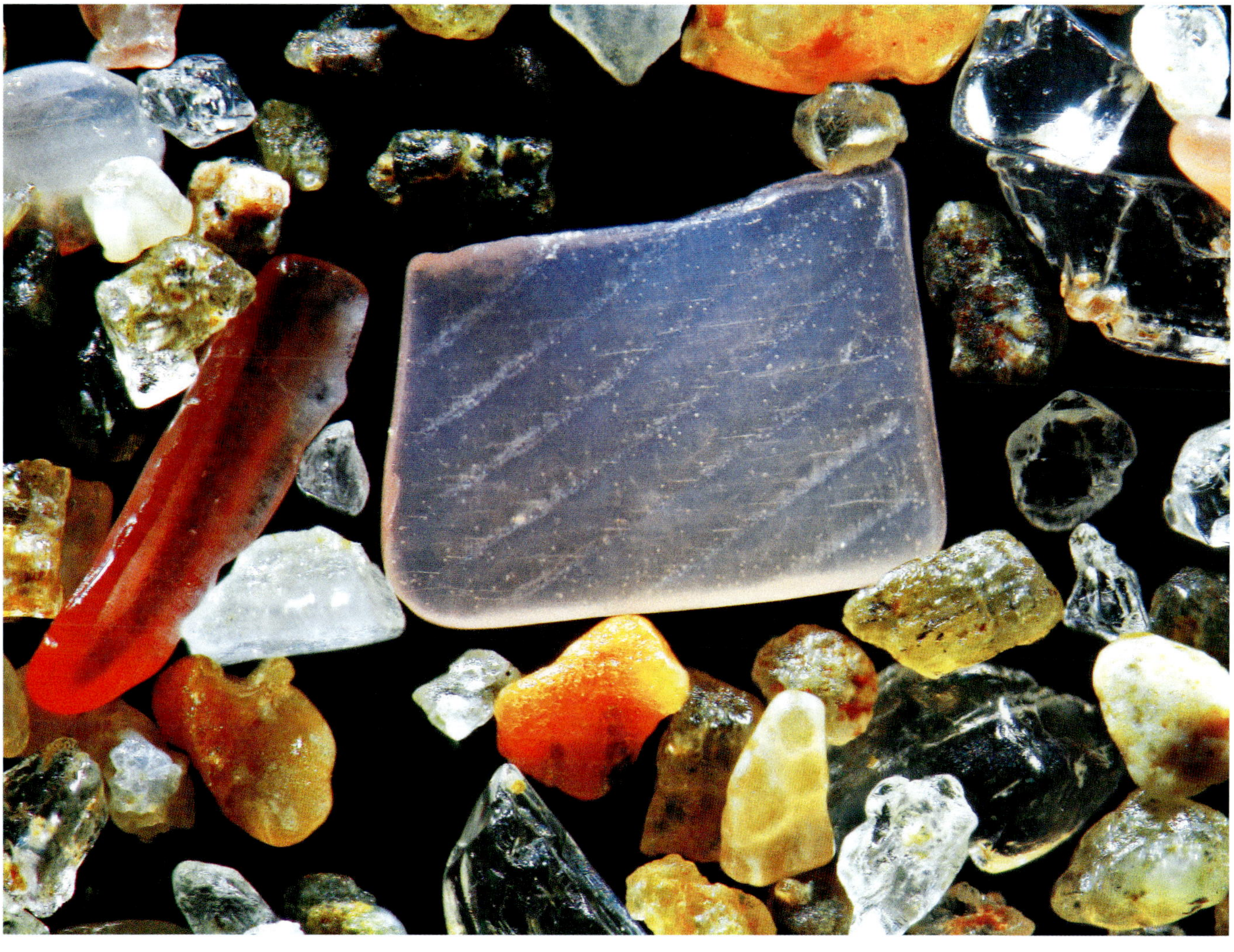

A square-shaped shell fragment is found amid mineral sand from Masaya, Nicaragua (magnification 85x).

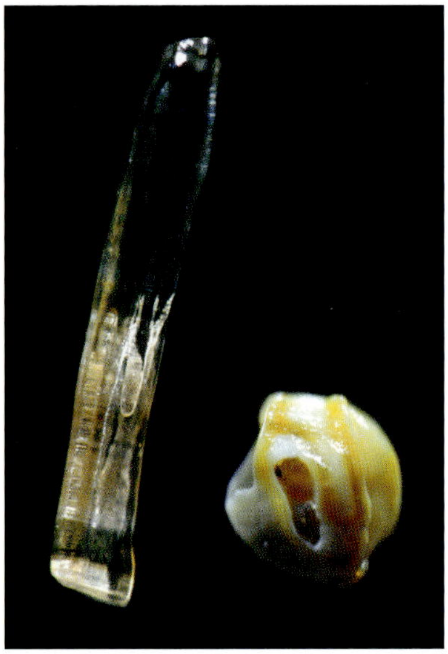

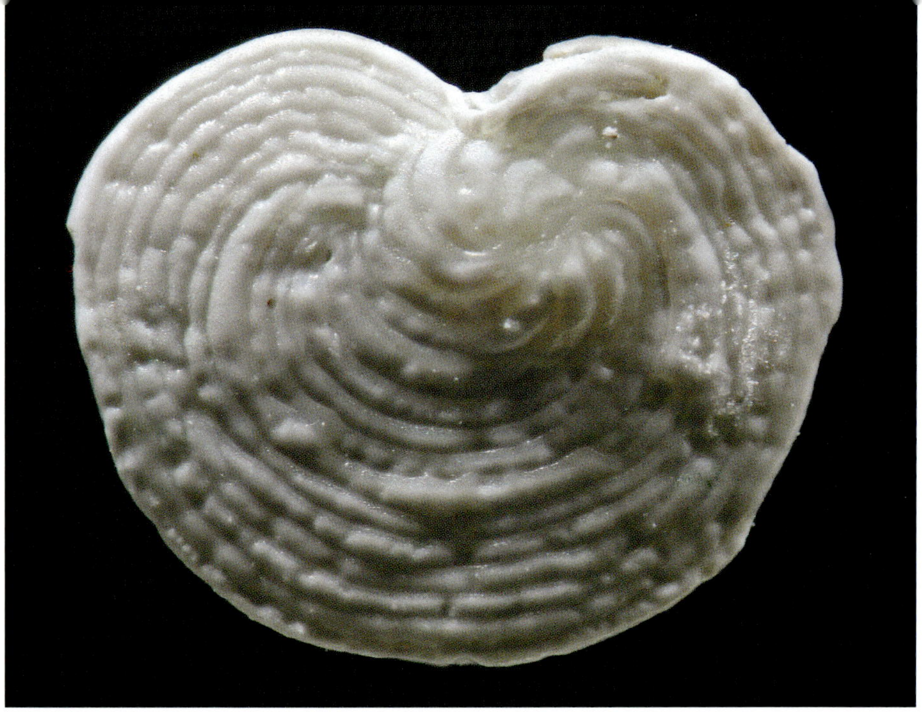

Top left: A pair of beautiful grains of sand are from Maui, Hawaii. The one on the left is a long mineral crystal and the other one is biogenic (magnification 70x). *Top right:* A heart-shaped grain of sand from Andros Island, Bahamas, is actually an intricate little foram (magnification 60x). *Opposite:* A baby sea urchin spine has become a grain of sand. Cross your eyes slightly to create a third middle image, which will give a 3-D view of the organization of the spine. Note the intricate internal structure, which is not visible in the flat 2-D image. The structure is made of crystals of magnesium-rich calcite that formed through self-assembly of molecular components (magnification 75x).

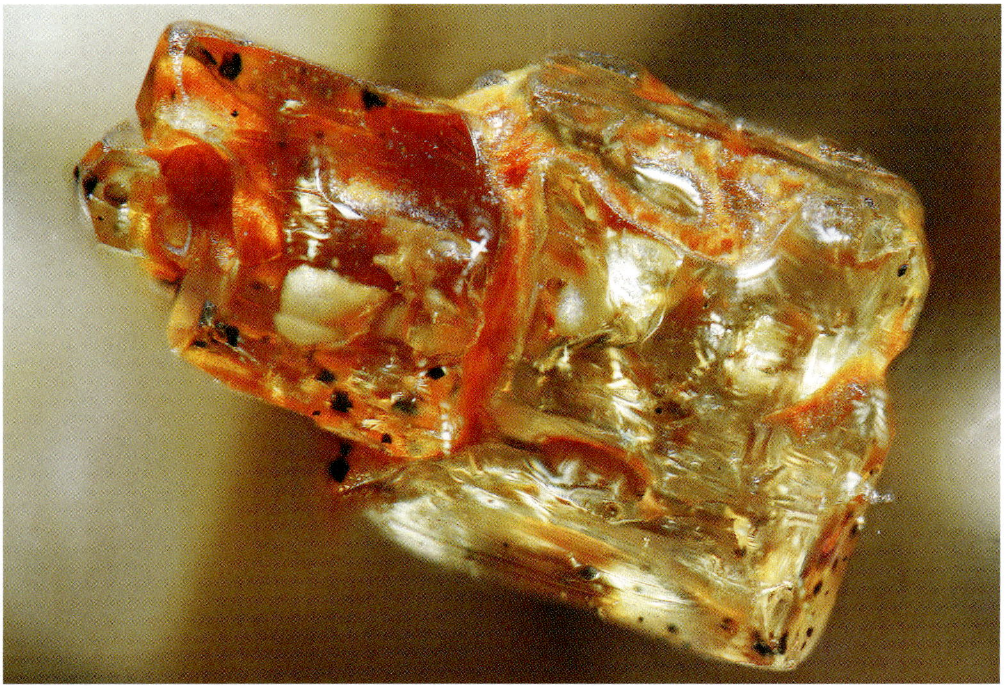

Sand from Sandy Bay, Oahu, Hawaii, contains a stunning chabazite crystal (magnification 155x).

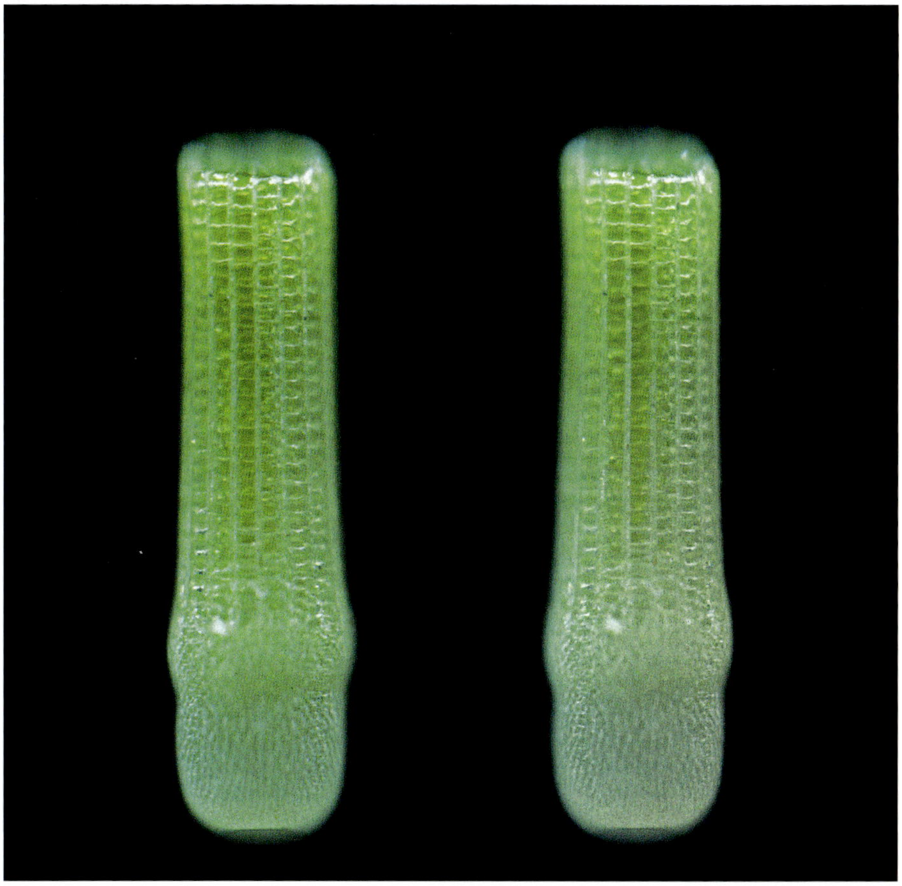

parts into complex mineralized shapes. This is one of the goals of nanotechnology, where useful microscopic objects are fashioned from molecular building blocks. The biological constituents of sand, such as a fragment of a sea urchin spine, stand as a testament to the beauty of nature's inventions. A stereo image reveals the wonderful 3-D structure of the sea urchin spine.

Many biogenic sand grains demonstrate recognizable geometry and symmetry, and these shapes and patterns aid in identifying the multitude of distinctive grains of sand on the world's beaches. Urchin spines are readily identifiable; their color may vary, but the shape remains the same whether from the cold North Atlantic or the tropical waters of Bermuda.

The Shapes of Biogenic Sand

The biological constituents of sand assume some of the most intriguing and intricate shapes of all. Evolution has created some spectacular mechanisms for fabricating calcified objects, such as shells, sponge spicules, and sea urchin spines. In a multifaceted process known as biomineralization, cells secrete a series of proteins that are involved in a sequence of chemical reactions. These chemical reactions lead to the self-assembly of calcium and other component molecules into mineralized tissues. It is the process of biomineralization that gives us the bones that enable us to walk upright. And it's the same process that endows us with teeth so we can chew our food.

Scientists are exploring the different ways organisms have solved the problem of self-assembly of small

Forams produce sand-sized calcium carbonate shells with chambers where unicellular organisms live. Some are globular in shape, while others have flattened or compressed shells. Forams can be found floating in the ocean or living on the sea floor; they also inhabit brackish mangrove swamps. Identifying the species of a foram is challenging, considering scientists have cataloged more than forty thousand, both living and extinct. Sensitive to pollution, forams function as bio-indicators of the health of an ecosystem.

Some biogenic grains are easy to identify because of their spiral pattern or growth rings. Other times, fragments are just that—fragments—and one must imagine the fragment as part of a whole. In addition, the erosion process destroys many of the recognizable characteristics, making identification even more difficult.

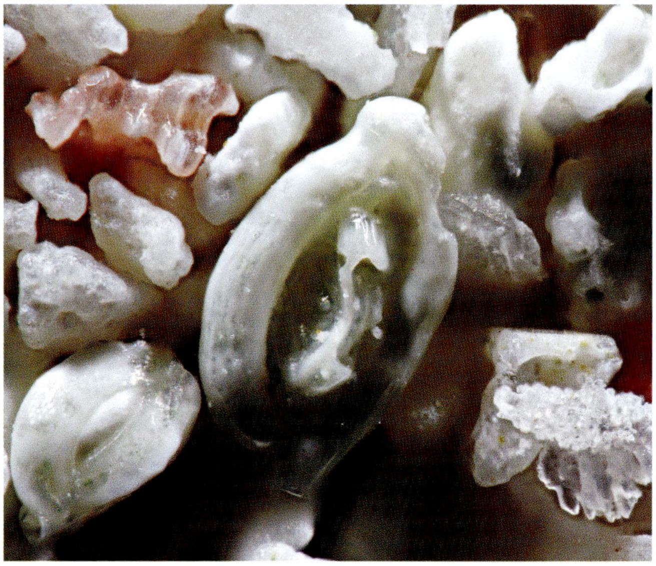

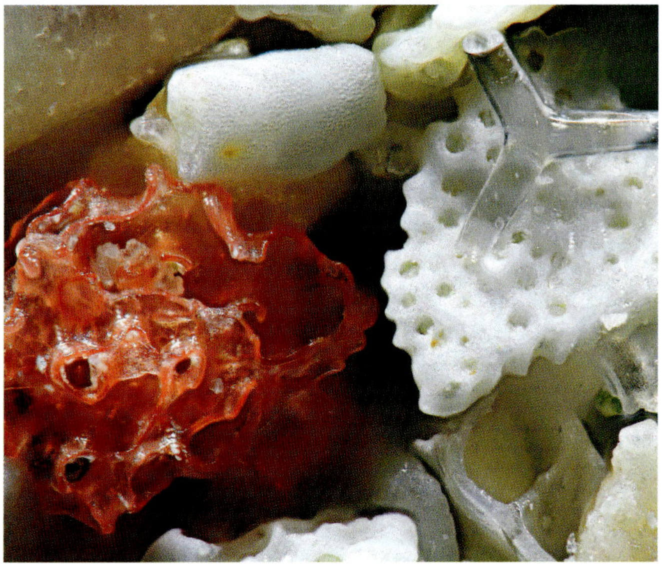

Forams make up much of the sand grains on Southampton Beach, Bermuda. *Left*: A porcelainlike foram takes center stage. *Right*: The tropical island sand contains a delicate silicate sponge spicule, a fragment of white coral, and a pink foram (magnification 100x).

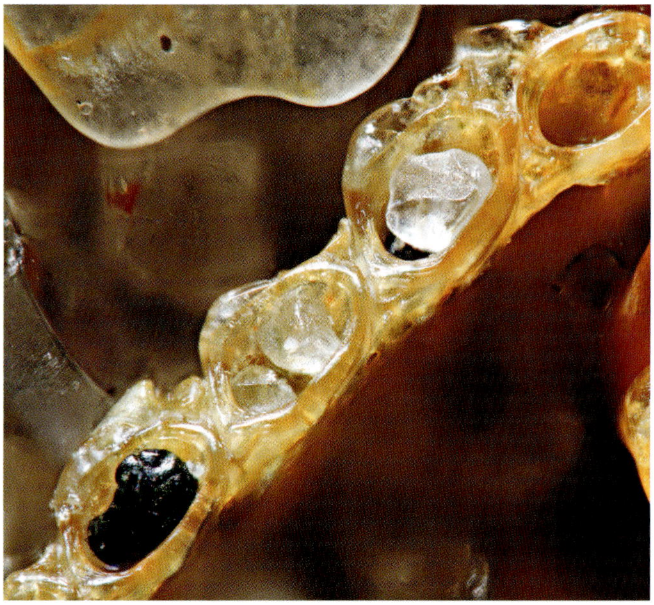

Tiny mineral sand grains sit in the hollows of a biogenic fragment from Geriba Beach, Brazil. The fragment looks like the jawbone of a tiny fish but is probably a bryozoan, a moss animal (magnification 80x).

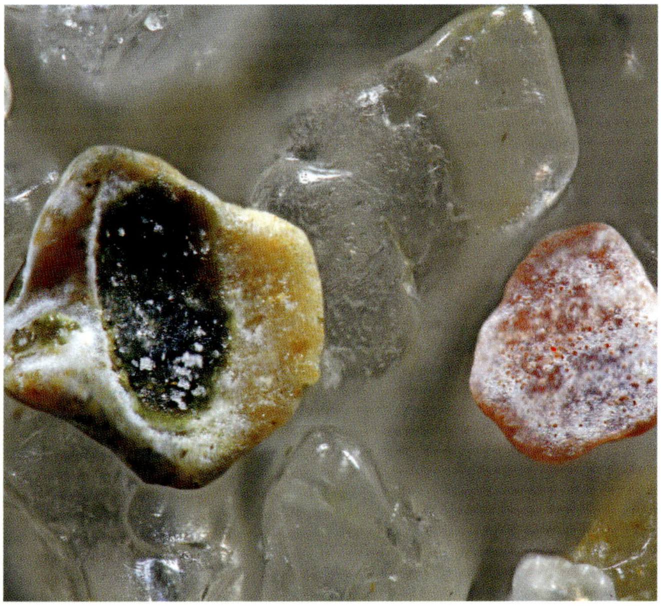

An image of sand from Copacabana Beach, Brazil, focuses on the deep interior of an eroded shell fragment surrounded by quartz sand grains (magnification 110x).

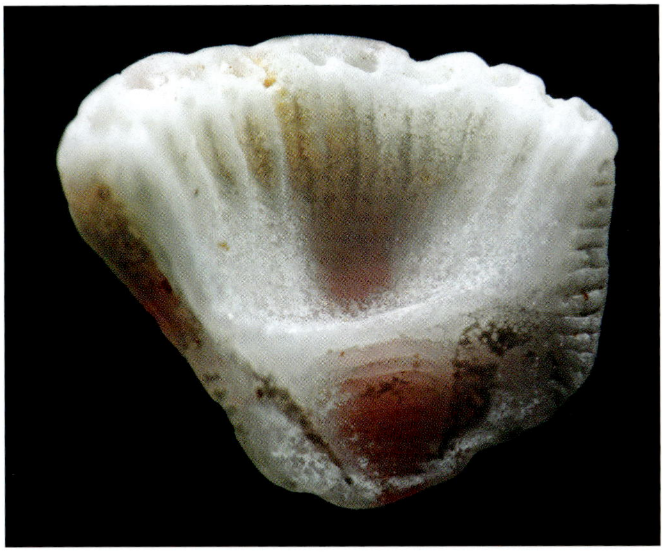

Sand from the Galápagos Islands illustrates the region's diverse ocean life. *Left*: A shell fragment is the hinge end of a tiny bivalve shell (magnification 105x). *Below*: A sample contains green-and-pink sea urchin spines and coral fragments (magnification 170x).

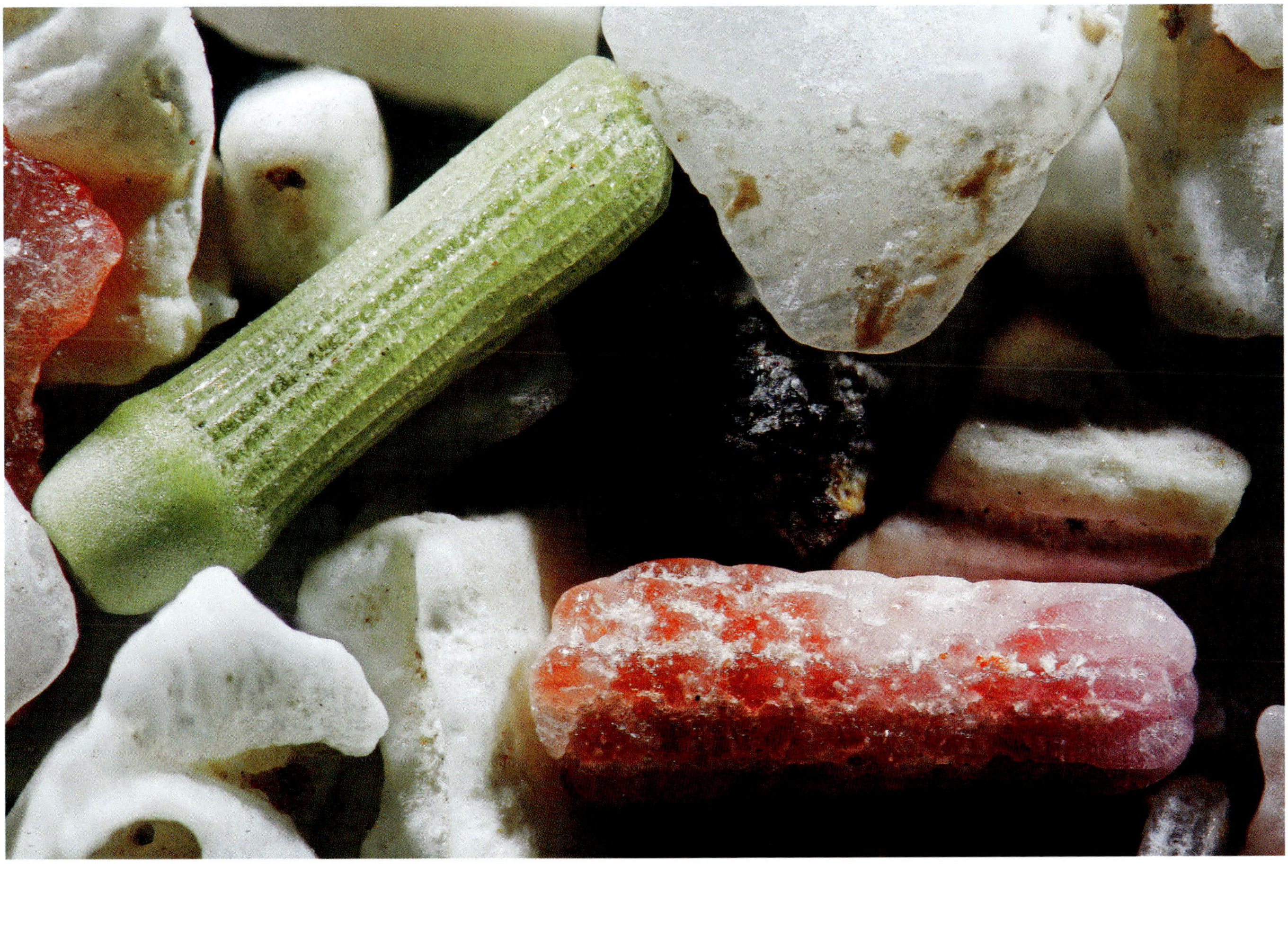

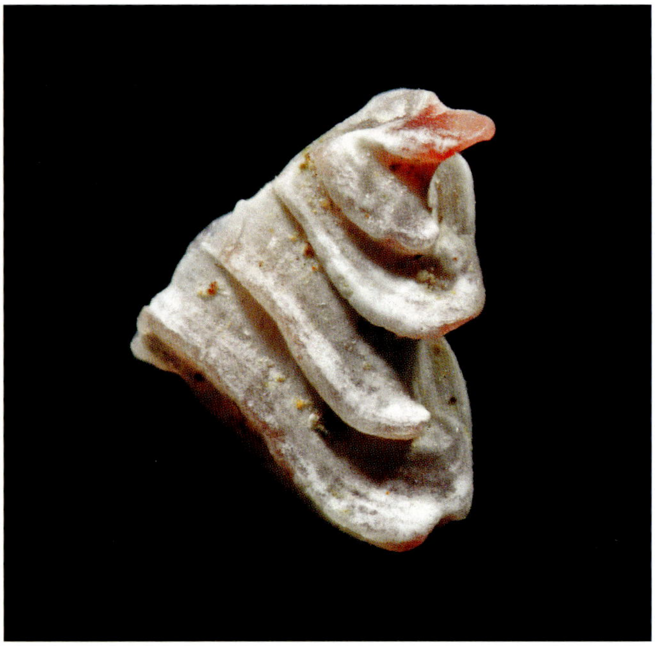
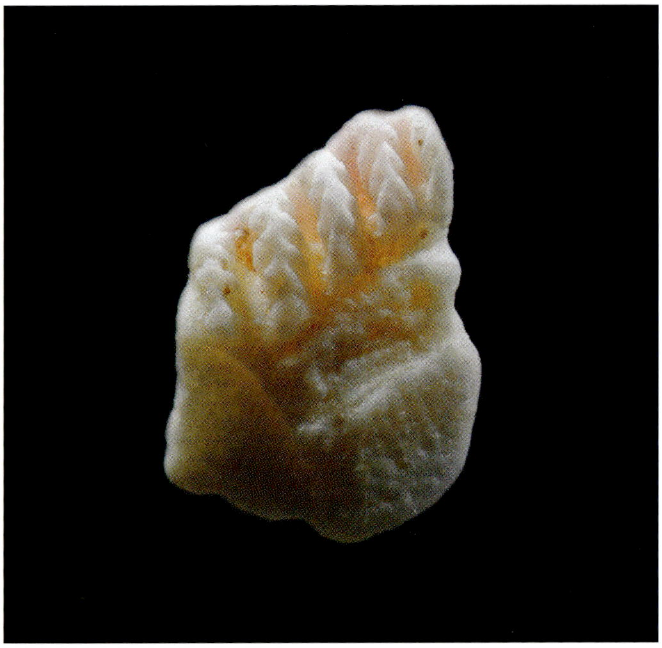

Sand grains from the Galápagos Islands take on the form of (*left*) a ridged shell fragment resembling frosting on a cake (magnification 90x) and (*right*) a beautiful multichambered foram (magnification 45x).

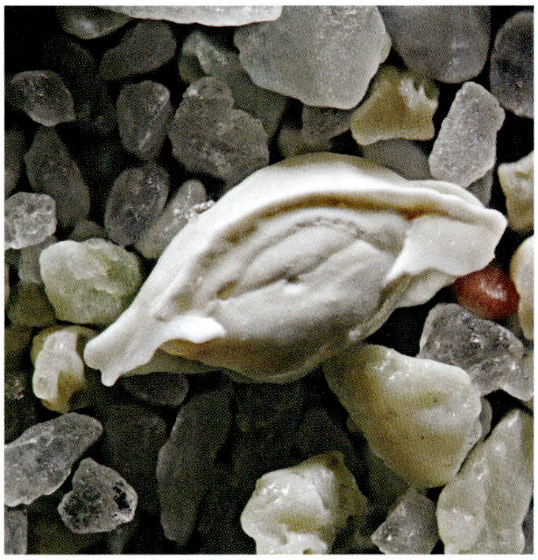
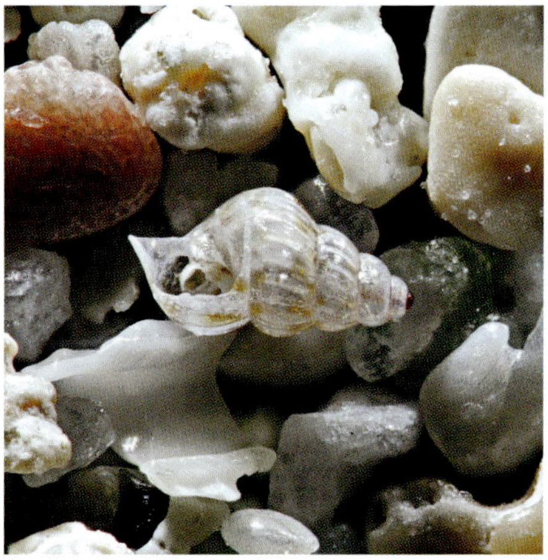
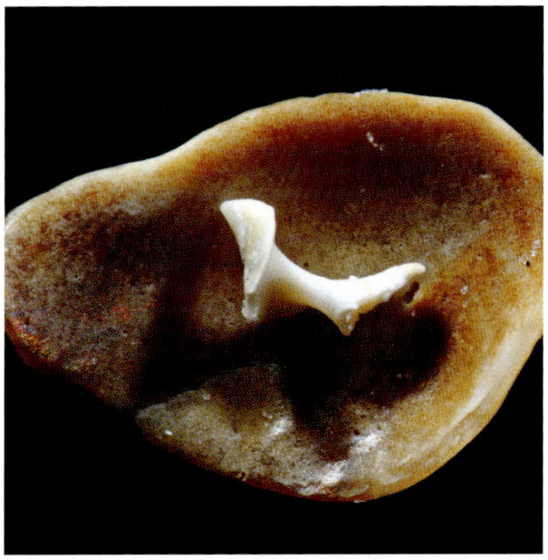

Sand grains from a beach on the island of Corsica, France, are varied. *Left*: A foram looks like a porcelain human eye. *Middle*: A perfect microscopic shell appears delicate, yet it is strong enough to survive the beating of waves. *Right*: A shell fragment cradles the internal spiral of another shell fragment that has eroded to almost nothing (magnification 75x).

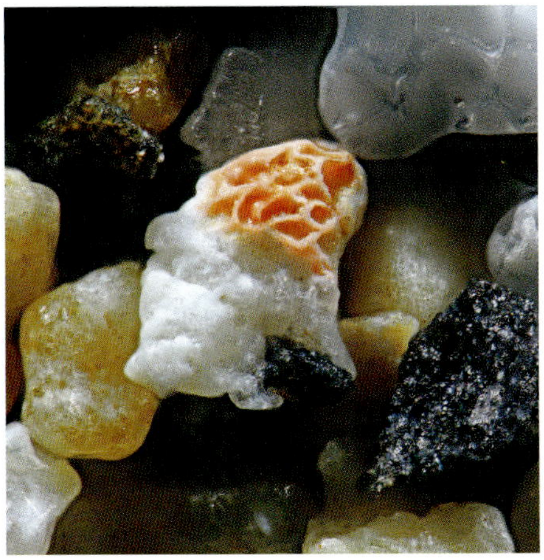

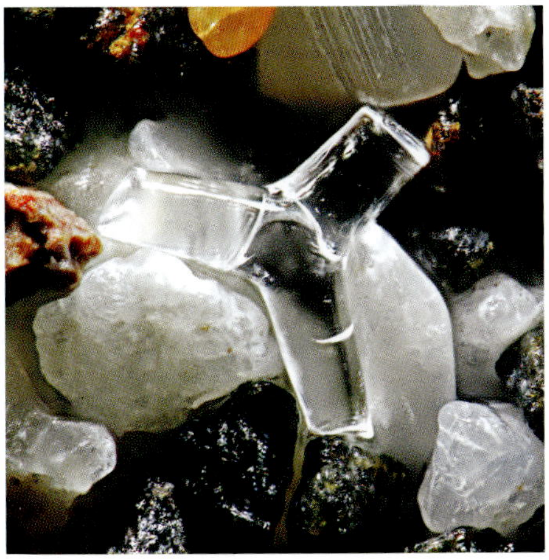

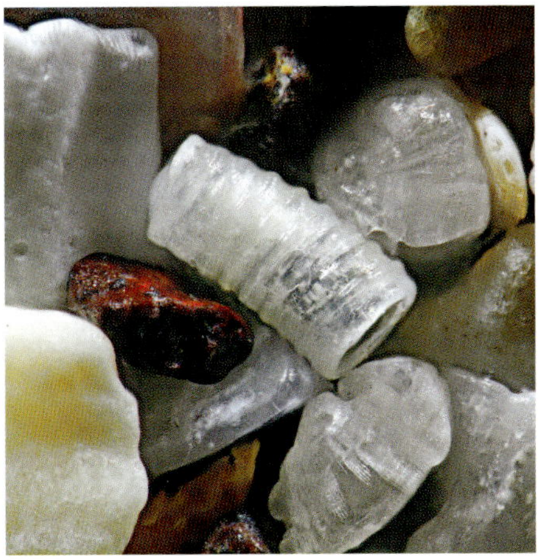

Sand from Hamoa Beach, Maui, Hawaii, contains (*left*) a magnificent piece of pink coral, (*middle*) a glistening sponge spicule, and (*right*) a tiny fragment of a tusk or worm shell (magnification 75x).

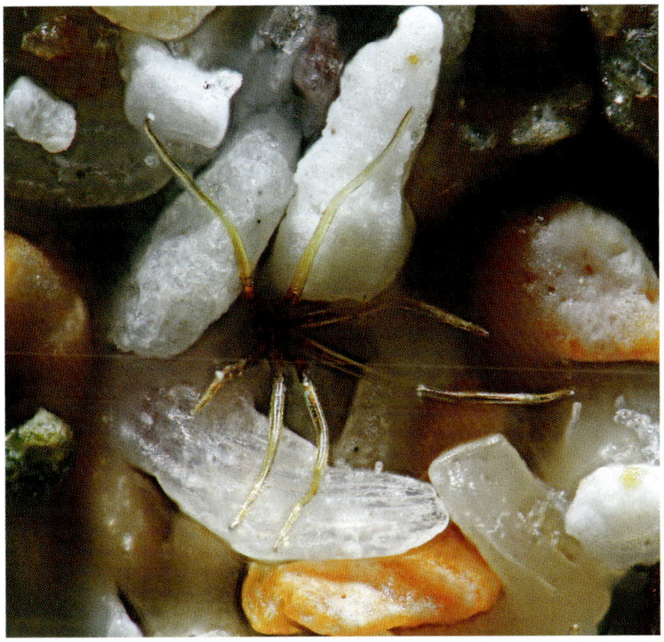

Sand from Tamarindo Beach, Guanacaste Province, Costa Rica, includes a delicate spiderlike sponge spicule. Siliceous sponge spicules occur in a variety of bizarre shapes resembling pick-up jacks, clubs, fishhooks, and spiny hairpins (magnification 115x).

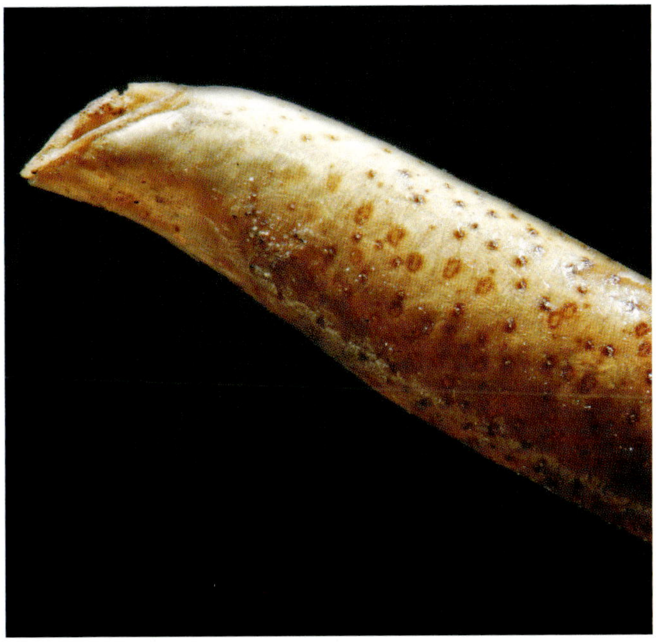

Sand from the Straits of Mackinac, Michigan, contains a marine fossil that is about 400 million years old. Tropical oceans once covered the region, and many marine fossils can be found. The fossil is perhaps an appendage segment from an arthropod (magnification 45x).

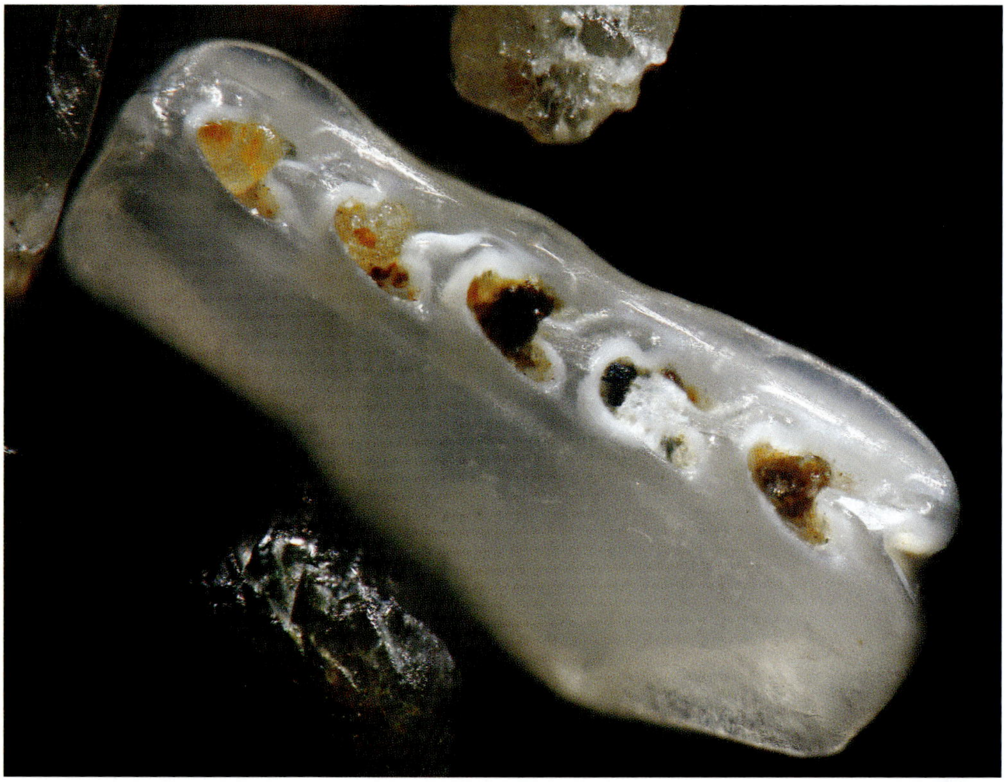

Tiny bits of minerals have embedded themselves into a shell fragment. The sand was collected from Masaya, Nicaragua (magnification 155x).

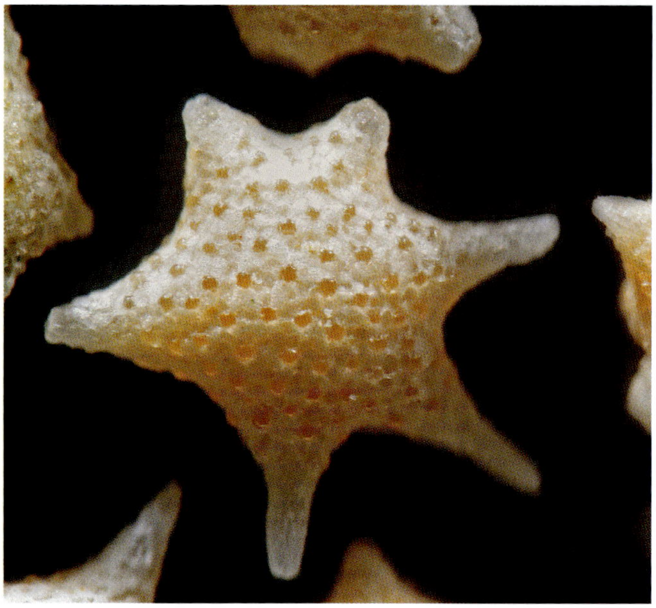

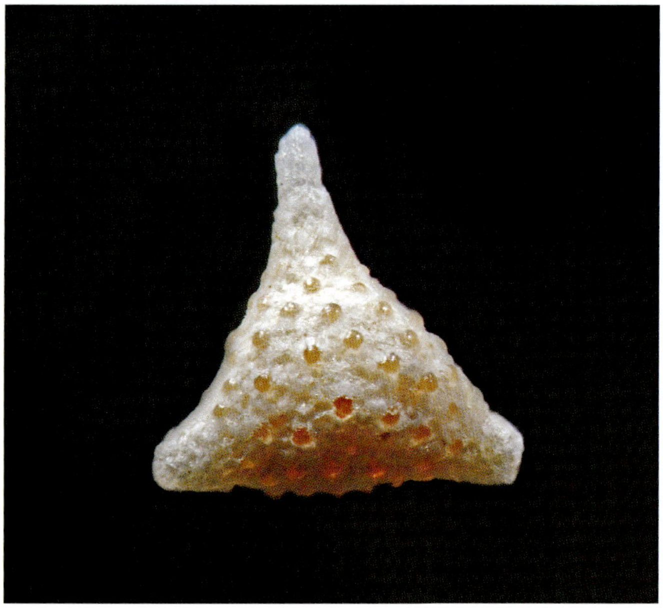

Sand from Taketomi Island, Okinawa, Japan, contains the region's famous star-shaped sand grains. The pearl-studded sand grains are forams, which commonly produce a calcified test, or skeleton, that is divided into chambers (magnification 45x).

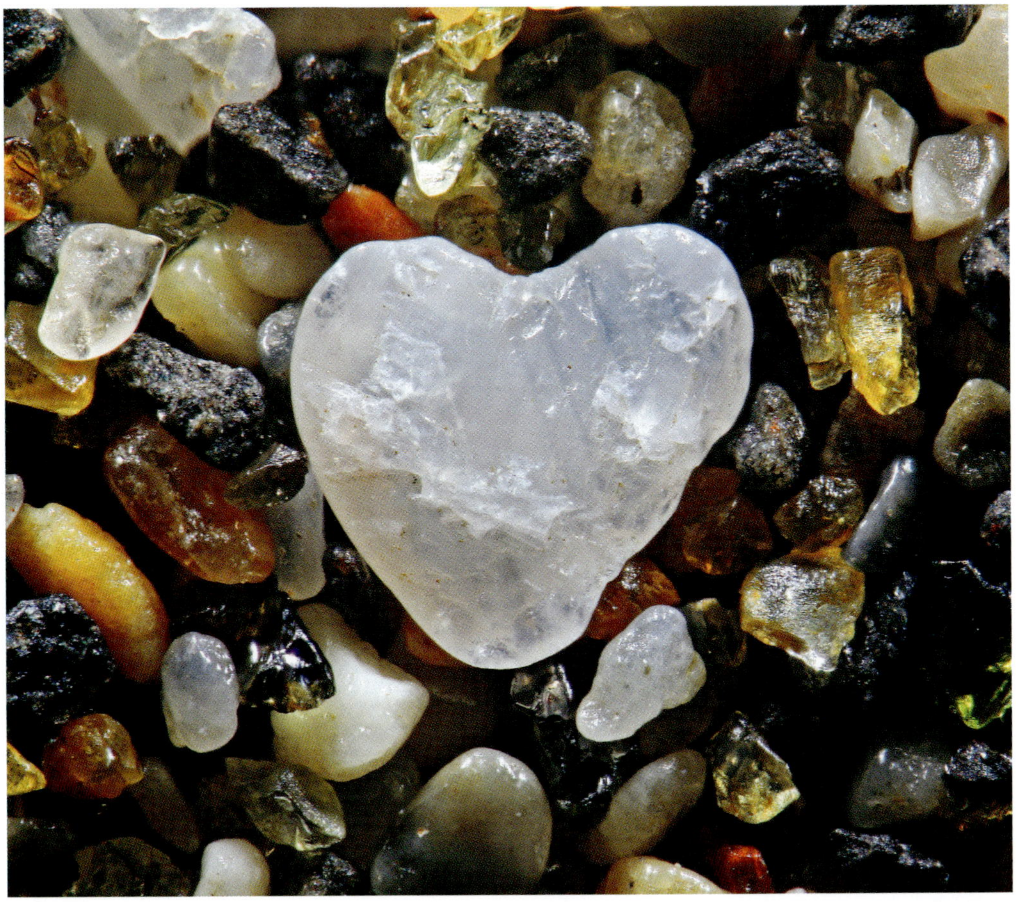

A beautiful heart-shaped shell fragment nestles among mineral sand from Lahaina, Maui, Hawaii (magnification 65x).

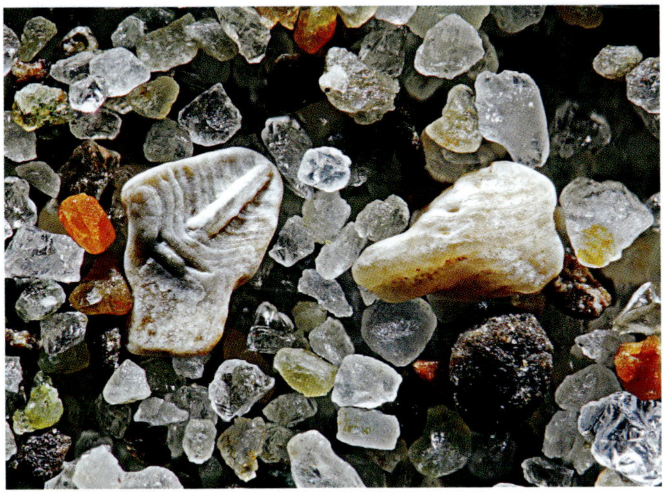

The internal cavity of a shell fragment is surrounded by quartz in a sample gathered from Stinson Beach, California (magnification 35x).

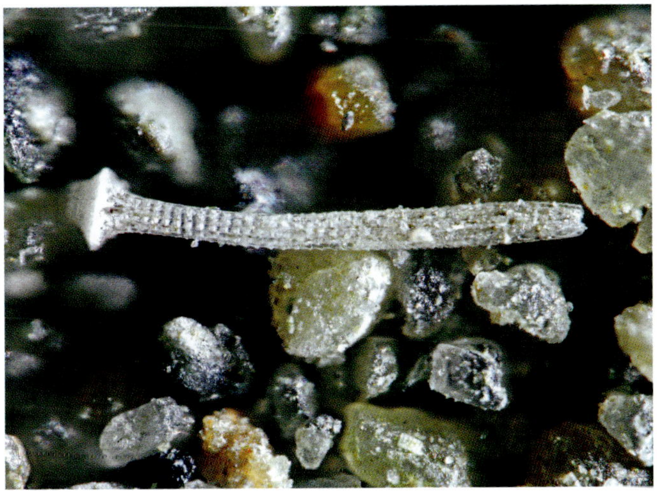

A delicate sea urchin spine looks like a nail in sand from Suquamish, Washington. The image shows the inflated base of the spine (magnification 80x).

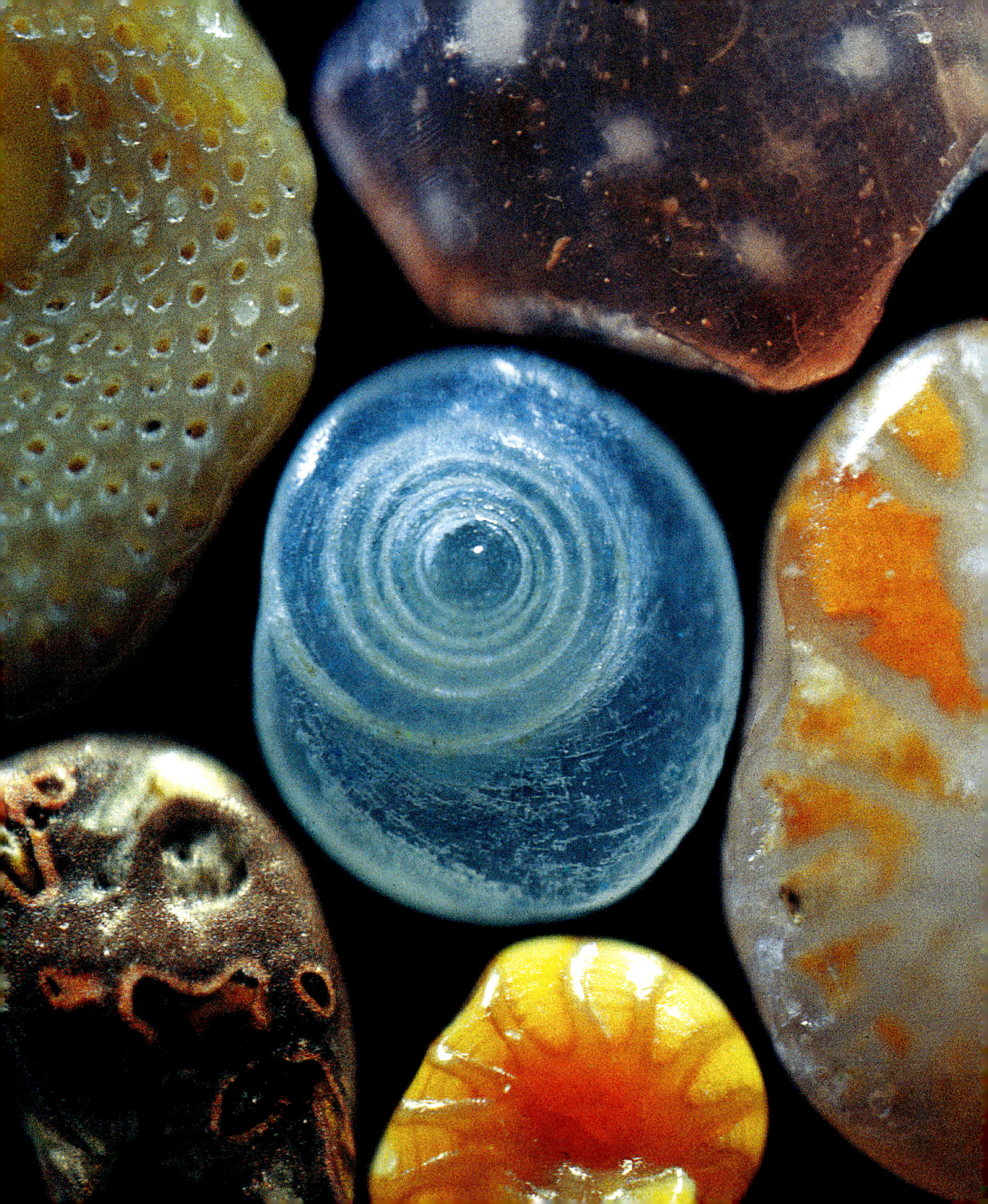

CHAPTER 6

Patterns in Sand Grains

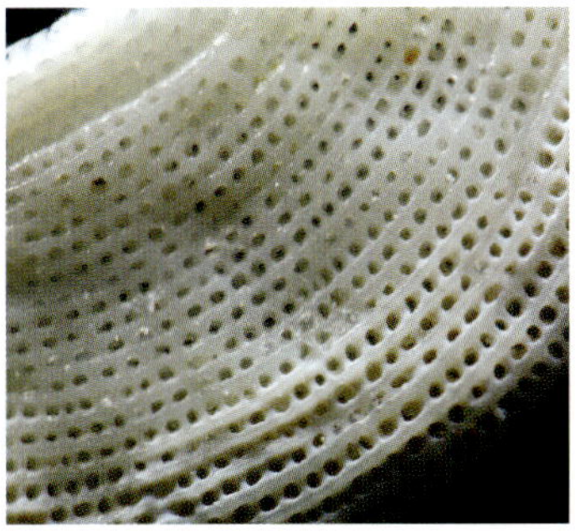 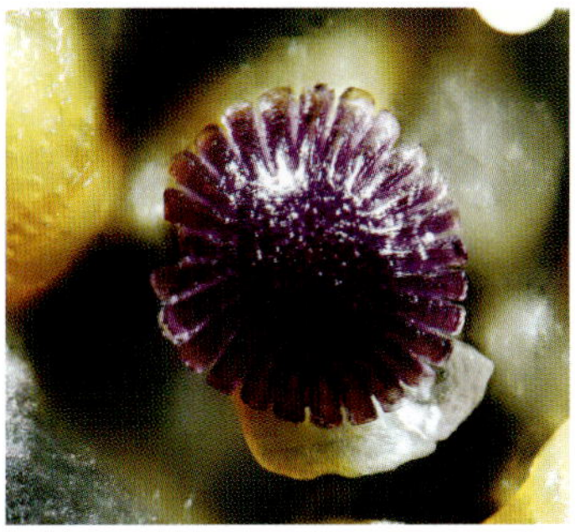 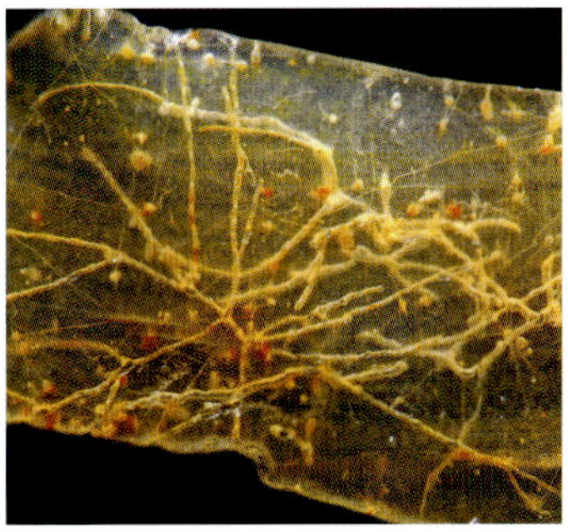

Looking at sand grains through the microscope is an awe-inspiring experience. To see the patterns that Mother Nature uses in her designs is truly inspirational. Gorgeous patterns and designs adorn different grains of sand from around the world, such as the sand on the beaches of Maui. Bold designs of spirals, polka dots, honeycombs, and stripes abound in nature, even on the microscopic scale.

Probably the most astounding patterns found in sand grains appear in fragments of sea urchin spines. The spines protect the animal's body from predators. In fact, venom from urchin spines can penetrate skin and be very painful. In addition, the spines protect the sea urchin from the effect of high-energy ocean waves.

Spines vary from species to species in color, shape, and microarchitecture. The color may vary from purple to green. More than seven hundred species of sea urchins exist in the world's oceans, the largest of which have spines twelve inches long.

The tip of a spiral shell has broken off and become a grain of sand. It is opalescent from the repeated tumbling action of the surf. Surrounding the shell fragment are five other sand grains, from top middle clockwise, (1) a pink shell fragment, (2) a foram, (3) a microscopic shell, (4) a volcanic melt, and (5) a bit of coral (magnification 225x).

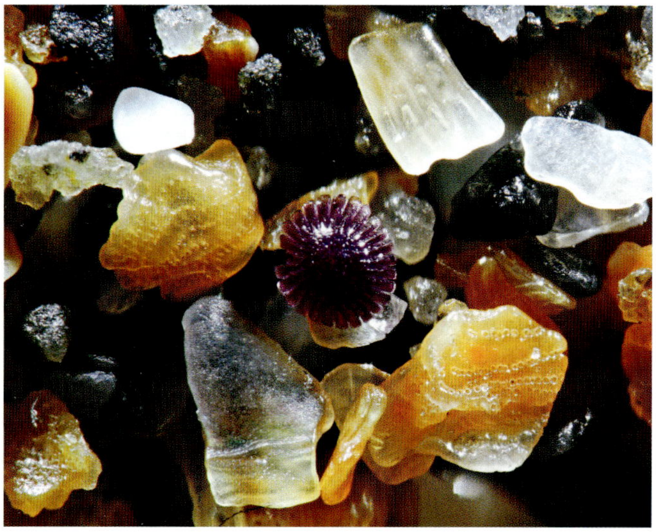

The Gulf Stream warms the waters around Fanore, Ireland, providing a suitable habitat for sea urchins. The bright-purple tip of a sea urchin spine is surrounded by shell fragments (magnification 35x).

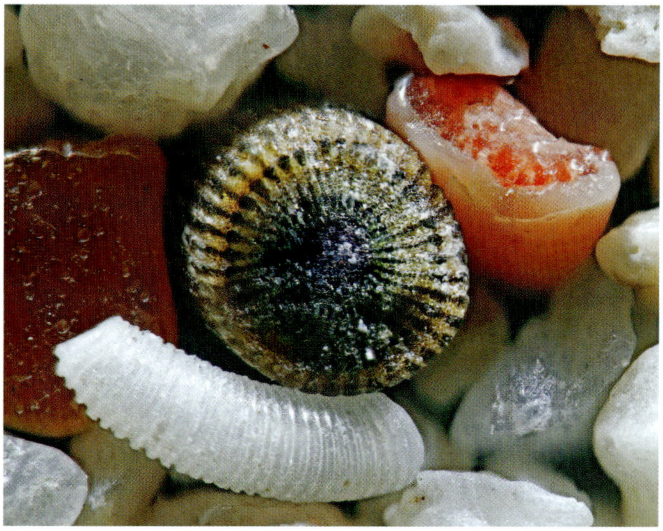

A sea urchin spine and tusk shell have distinctive designs in sand from the tropical waters of Utila, Honduras (magnification 100x).

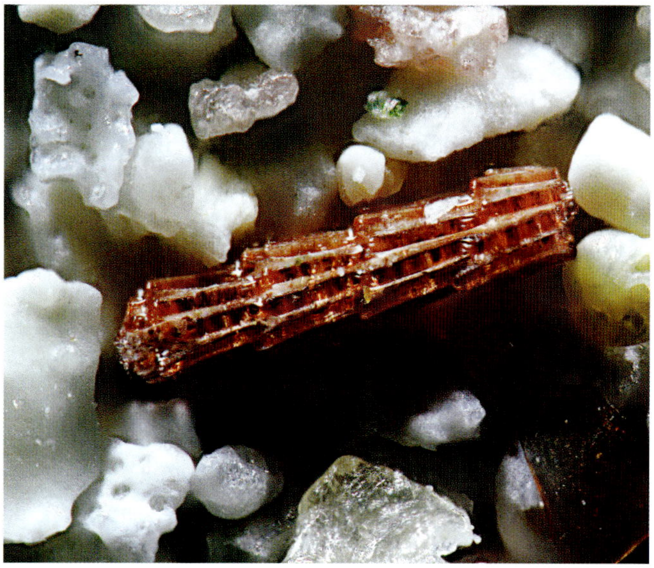

Sand from Southampton Beach, Bermuda, is predominately made of bits of coral, plus a pink urchin spine (magnification 70x).

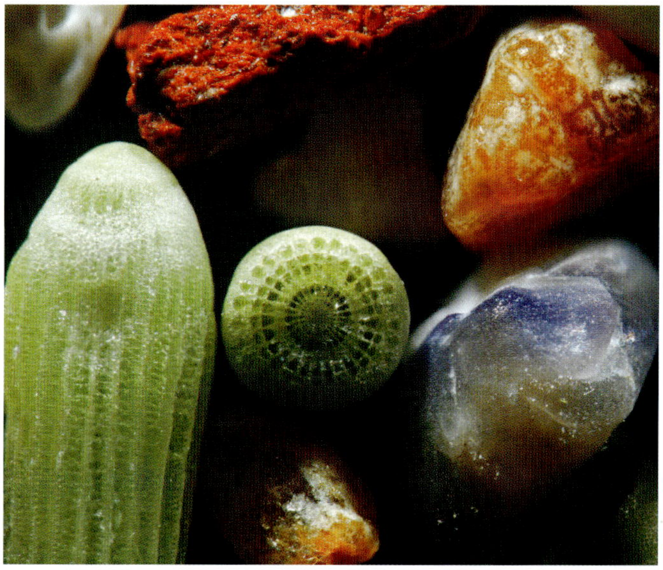

Sand from the Isle of Shoals, Maine, shows a mandala pattern in a cross-sectional view of a green sea urchin spine, alongside a spine seen in side view. The bricklike grain at the top of frame is staurolite and the blue mineral grain is kyanite, both metamorphic minerals (magnification 100x).

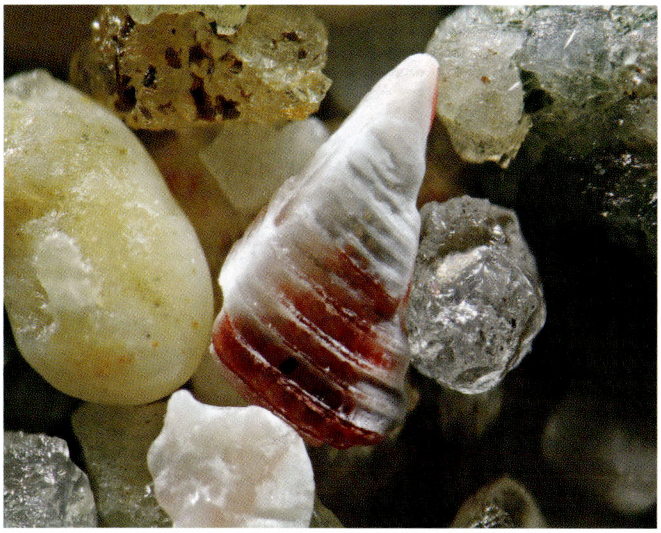

A striped shell fragment, an oval feldspar grain, and clear quartz grains compose a sand sample from Baja California, Mexico (magnification 100x).

Patterns in Mineral Sand Grains

Minerals found in sand on a particular beach tend to display similar characteristics of shape, degree of roundedness, and color. These features help identify the grains and trace the sand to its origins. Well-rounded grains such as quartz, which is chemically stable and physically hard, tell the story of millions of years of erosion and transport. Angular grains often indicate more recent erosion from bedrock and less transport; this history is also usually true of sand that contains many kinds of minerals.

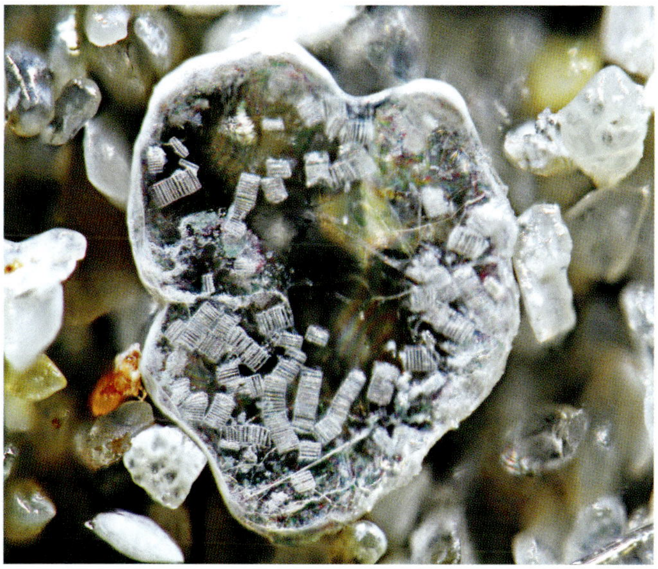

A grain of sand from Cable Beach, Australia, shows a mysteriously interesting pattern (magnification 100x).

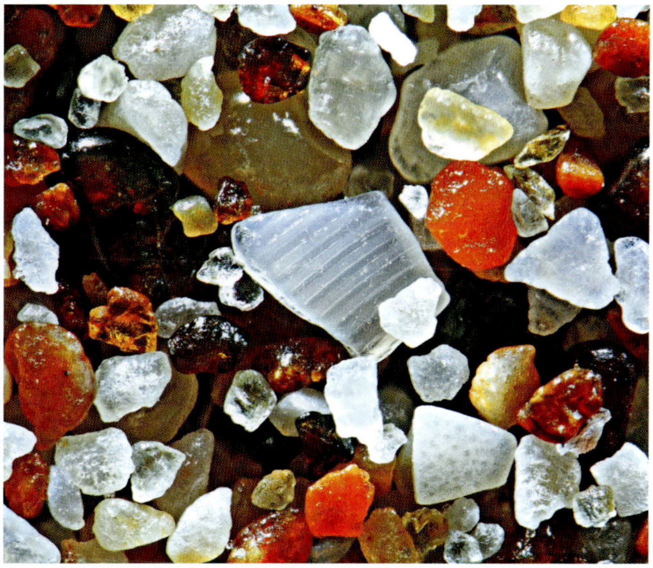

Sand from Dibba Beach in the United Arab Emirates contains a variety of mineral grains and interesting shell fragments. The minerals in the sand indicate it originated as continental rock (magnification 75x).

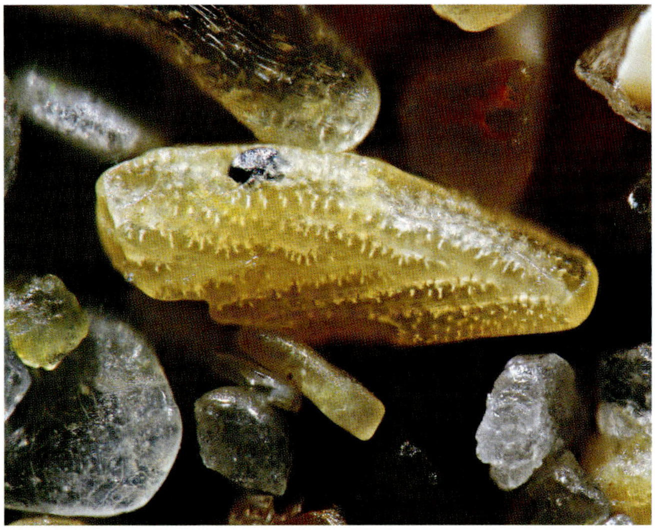

Sand from Homewood Beach, Eastern Cape, South Africa, on the Indian Ocean reflects the regional continental geology and marine life of the area (magnification 100x).

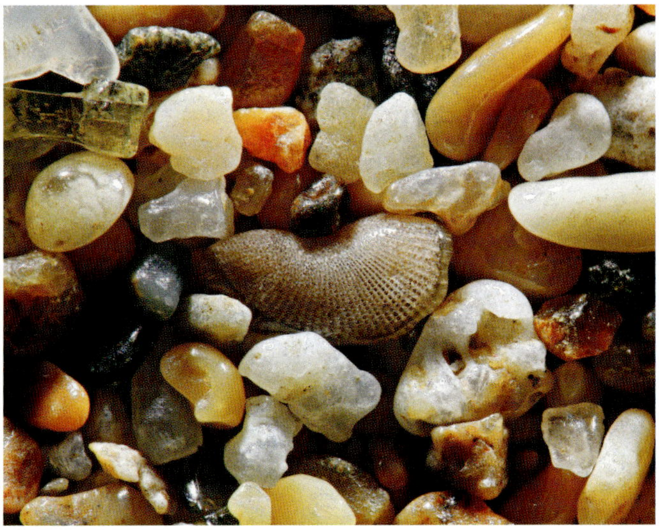

The diversity among sand grains from Kapalua, Maui, illustrates a healthy marine environment. The sample contains forams, shell fragments, and bits of coral, along with the igneous minerals chabazite and olivine (magnification 35x).

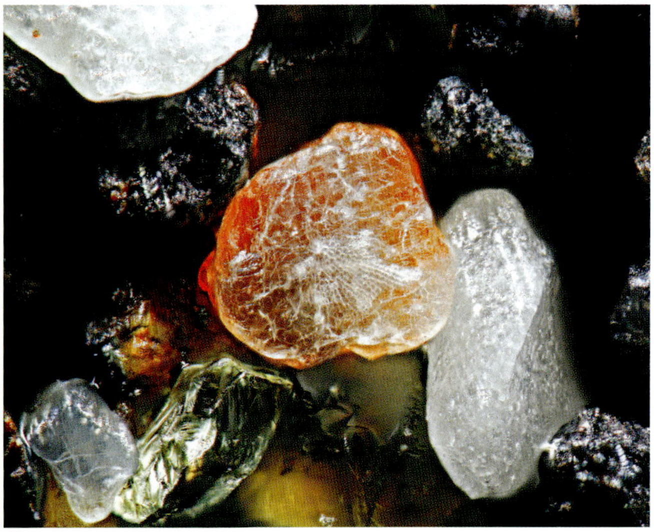

Sand from Kona, Hawaii, is predominately made of dark volcanic minerals, including small pieces of green olivine. A pink shell fragment adds some variety (magnification 100x).

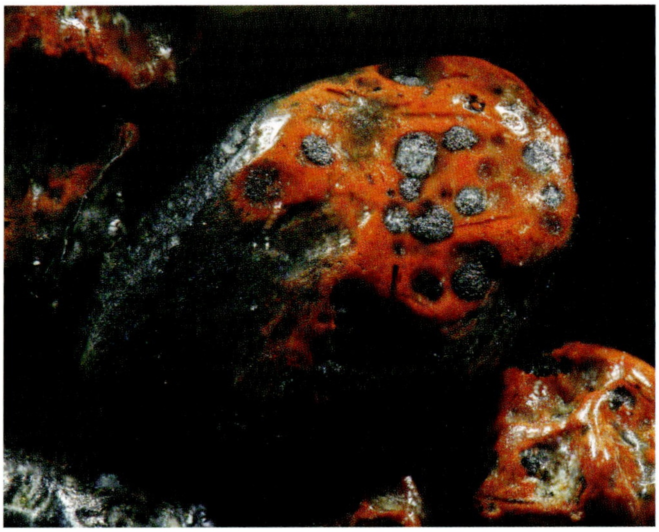

Glassy red-and-black volcanic grains are found in sand from Makena Point, Maui, and elsewhere around the Hawaiian Islands (magnification 100x).

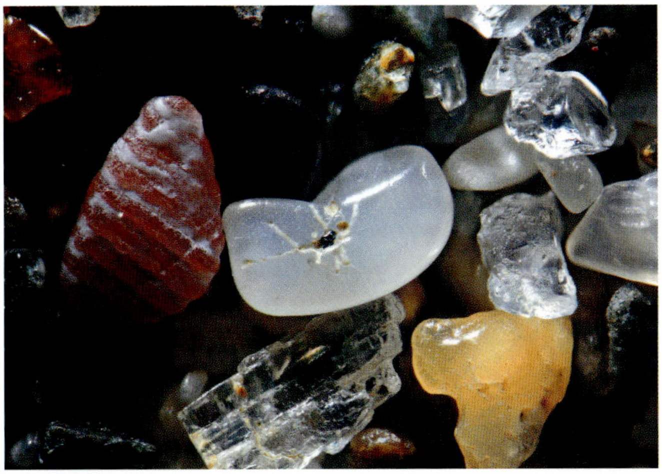

A couple of shell fragments can be seen in the center of a picture of sand from Masaya, Nicaragua. The middle fragment features worm tracks; just below it is a piece of clear feldspar; clear quartz is in the upper right (magnification 40x).

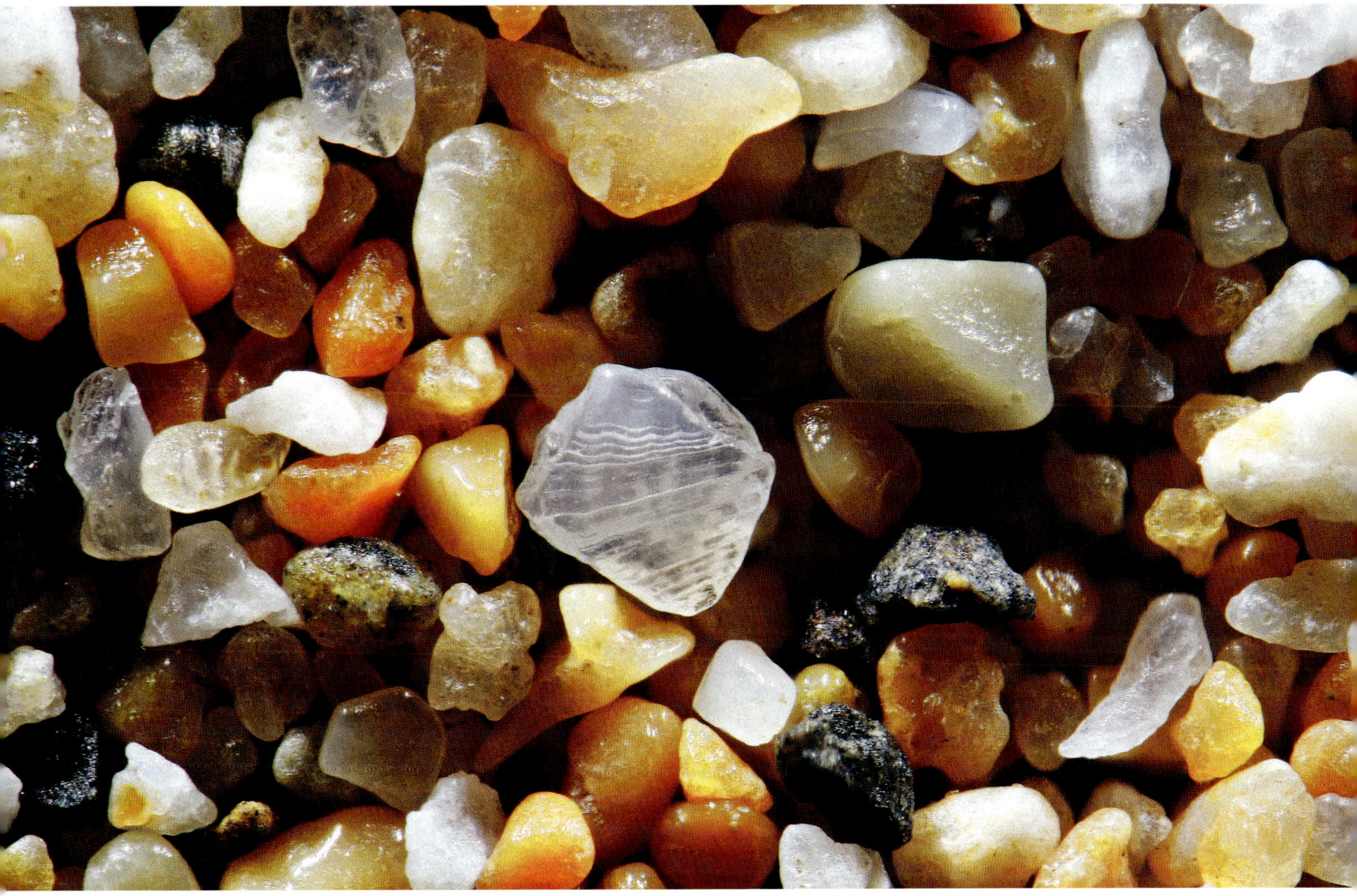

Sand from Ulua Beach, Maui, contains a wonderful mix of organic and inorganic components (magnification 85x).

Patterns in Biogenic Sand Grains

Spiral and mandala patterns are repeating themes in sand grains. Intricate mandalas occur in bits of sea urchin spine. It is delightful that nature has inscribed such beautiful patterns in tiny grains of sand.

The patterns are diverse and often of dazzling beauty. The emergence of pigment patterns on shells is an interesting case of biological pattern formation. A shell can only increase its size through accretion of new material along its marginal zone, which is the growing edge of the shell. Pigment is incorporated during growth at the leading edge, producing beautiful patterns on the surface of the growing shell. In fact, the patterns found on shells document the history of this dynamic process. This process is similar to the way patterns develop as textile is woven on a loom. Small changes in the way color is introduced can make big changes in the overall pattern being fashioned.

When we look closely at grains of sand through the microscope, we see art forms in nature. Sand allows us to appreciate the multitude of structures and subtle details that nature generates, such as the fabulous honeycomb pattern on the forams from beaches in the Bahamas and the radial symmetry of the cross section of an urchin spine. The growth rings on shell fragments and the internal anatomy of shells are both outstanding examples of Mother Nature's creativity.

After death, biological species become part of the debris on the ocean floor. Their calcium carbonate shells slowly break down. Sometimes, biological organisms help the process of decomposition. A shell fragment from Geriba Beach, Brazil, shows worm trails through the shell, and beautiful fragments in sand from Utila, Honduras, are pitted from the chemical reactions of the ocean water.

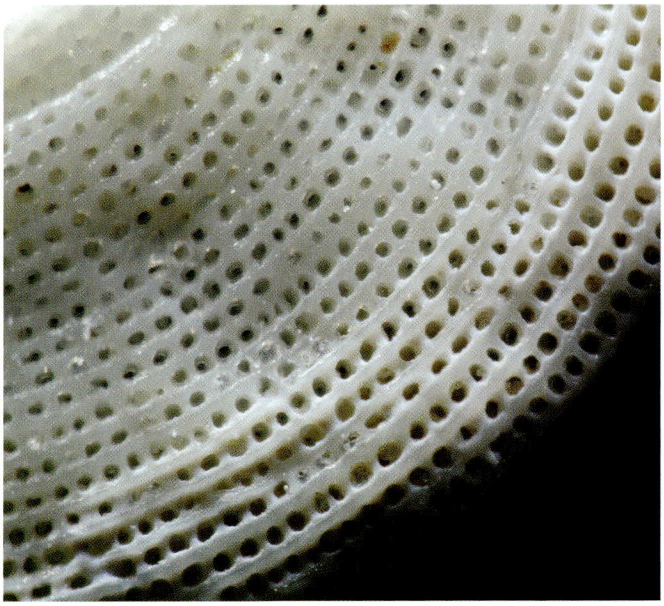

A beautiful close-up of a foram from Andros Island, Bahamas, shows the intricate microstructure of the species (magnification 110x).

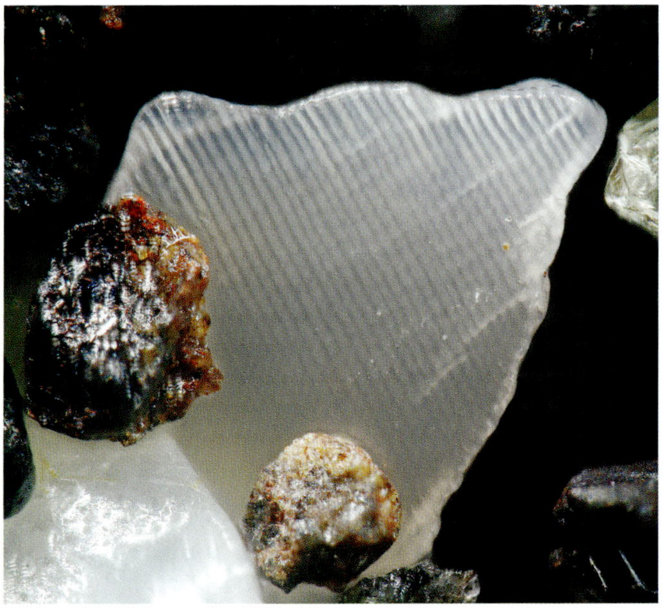

The pattern of growth rings can be clearly seen in a tiny shell fragment collected from Kona, Hawaii (magnification 100x).

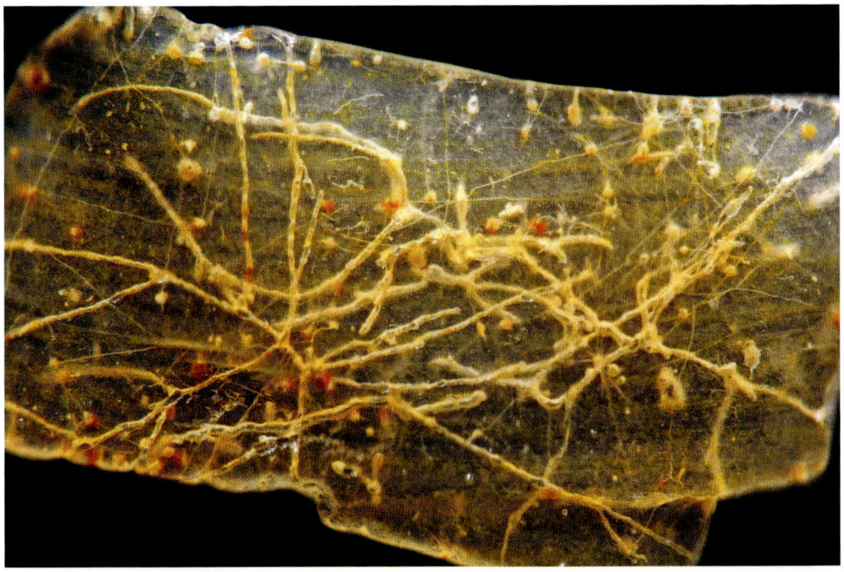

Sand from Geriba Beach, Brazil, contains a shell fragment bearing microscopic worm trails (magnification 110x).

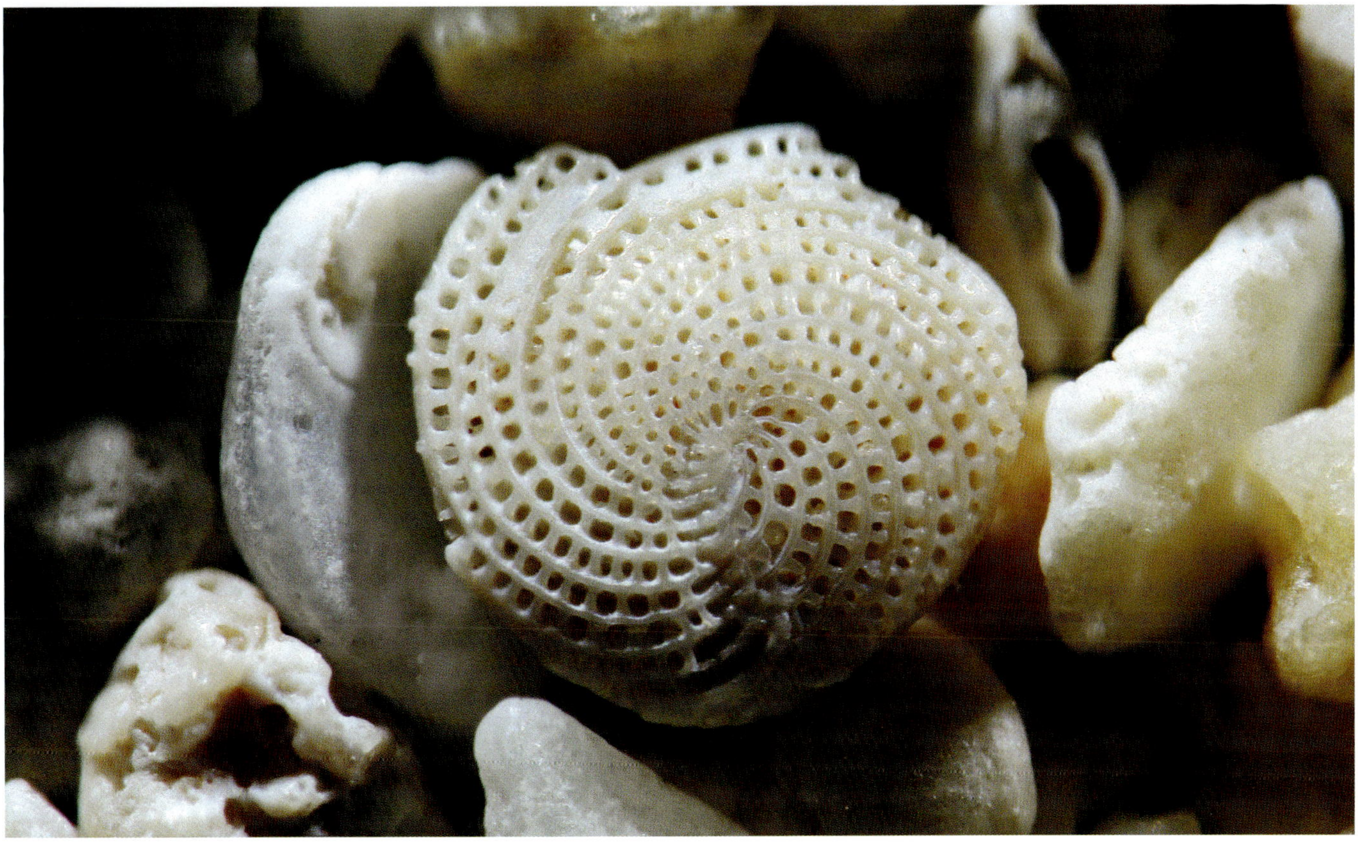

A beautiful foram from Utila, Honduras. This sand has an intriguing spiral honeycomb design (magnification 210x).

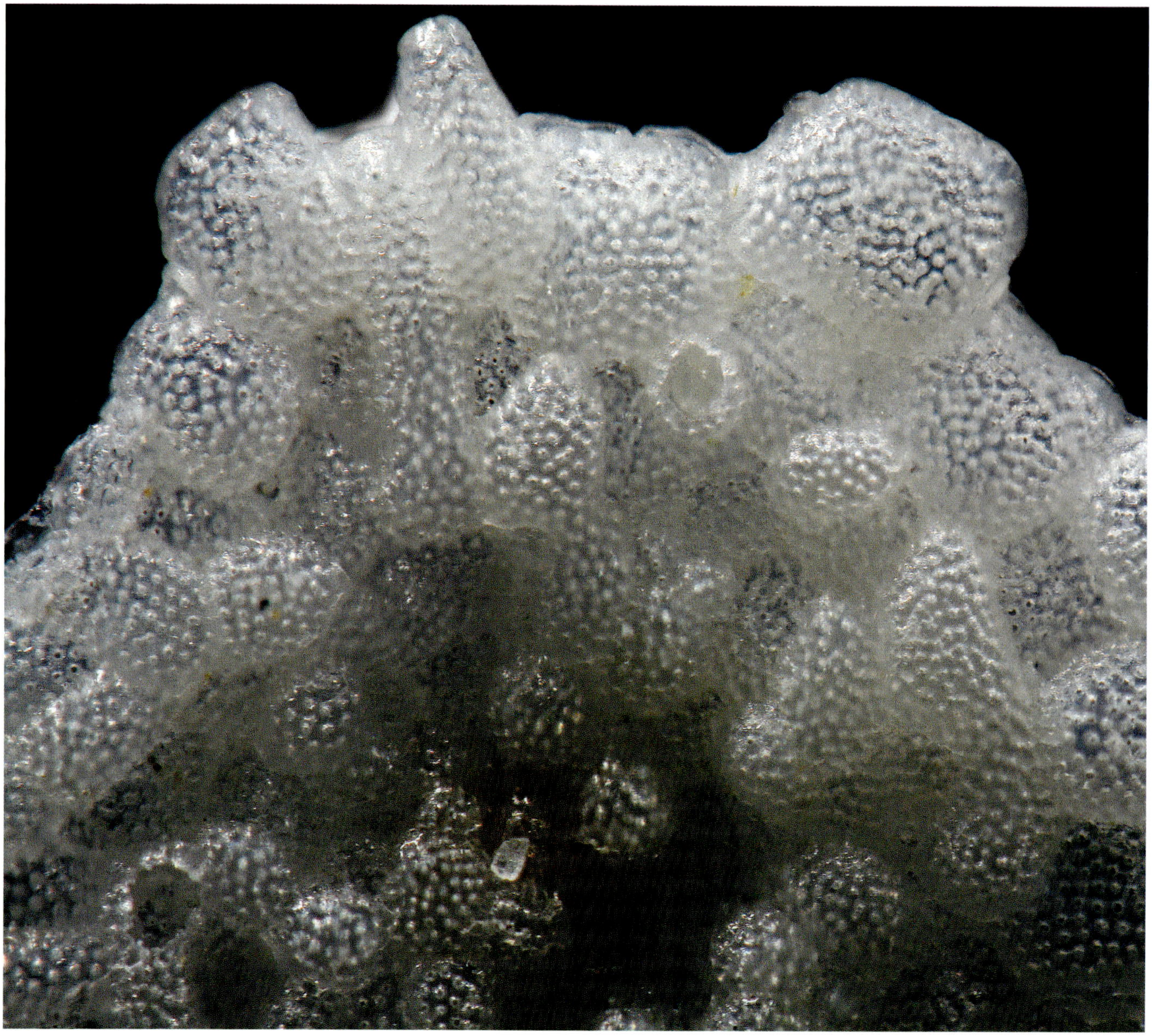

Sand from Andros Island, Bahamas, contains a delicate foram. Yet it is robust enough to tolerate the abuse of the ocean waves and end up on the beach as a grain of sand (magnification 215x).

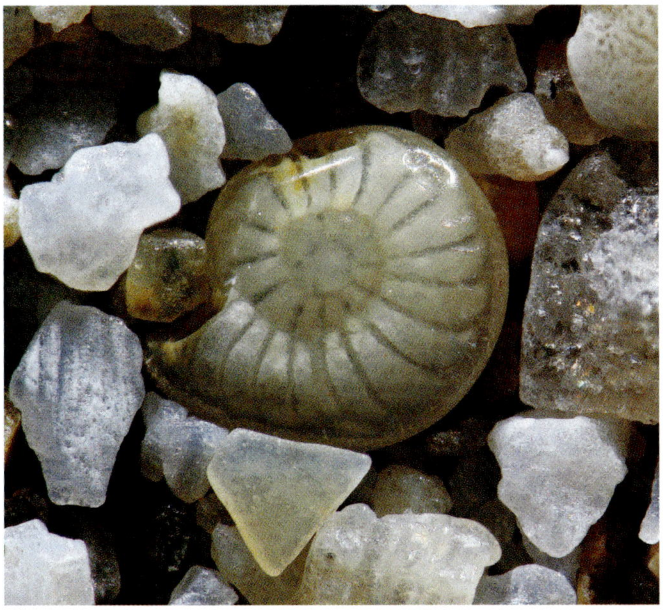

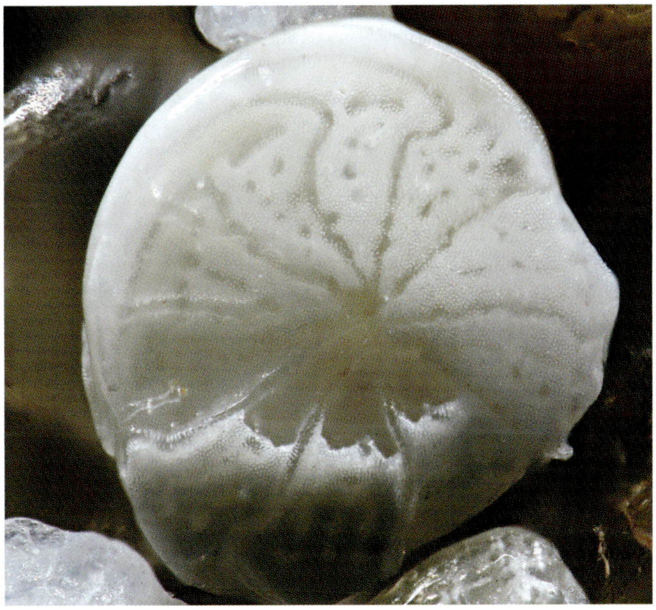

Beautiful forams are found in sand from Hainan Island, China. Forams exist in all marine environments and may live in shallow or deep waters. Many forams have chambers arranged in spiral patterns (*left:* magnification 45x; *right:* magnification 100x).

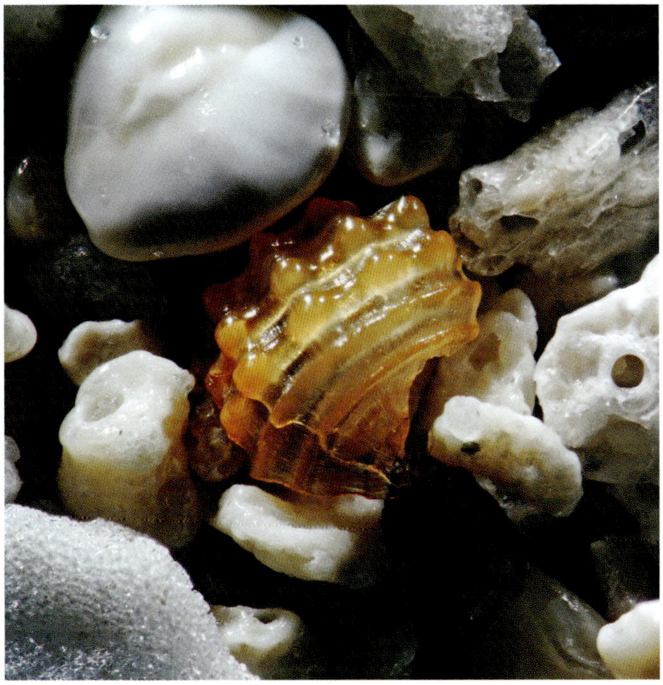

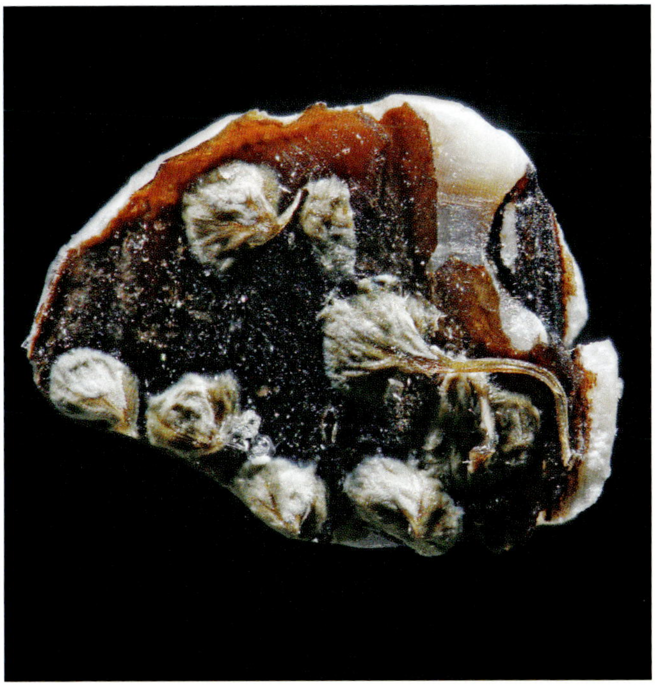

Left: Sand from the island of Corsica, France, contains a fragment of a tiny gastropod shell. *Right*: An unidentified grain of sand from the Isle of Shoals, Maine, is extremely unusual-looking (magnification 100x).

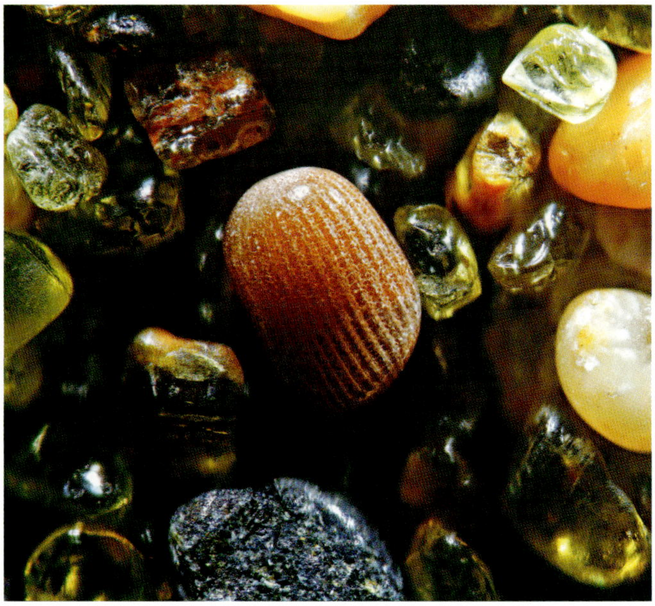

Sand from Lumahai Beach, Kauai, contains an elegant piece of pink sea urchin spine and wave-polished green olivine (magnification 100x).

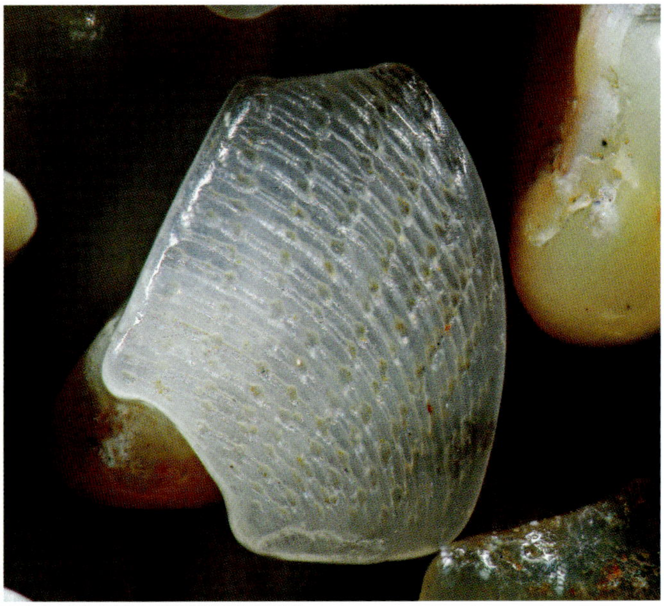

The radial ribs and grooves of a shell fragment can be seen among polished sand grains from Kapalua, Maui (magnification 125x).

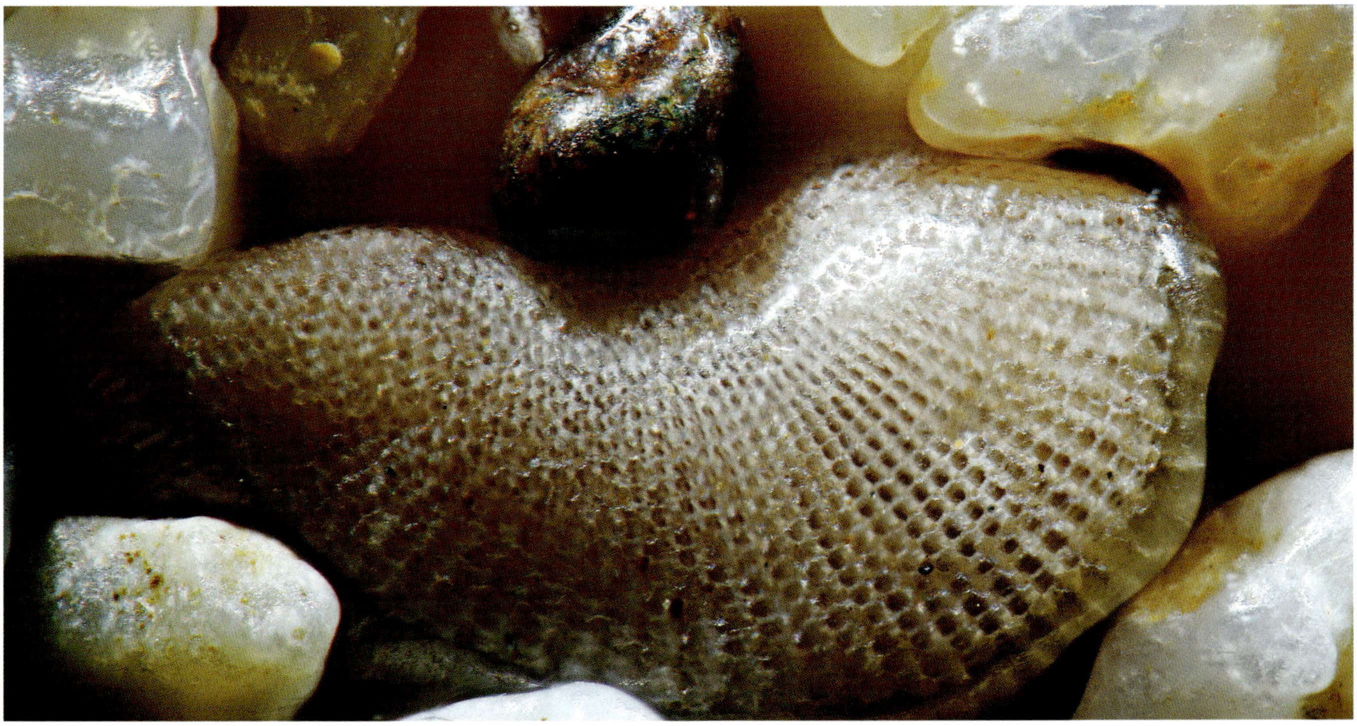

Radial patterning distinguishes a foram fragment found in sand from Kapalua, Maui (magnification 170x).

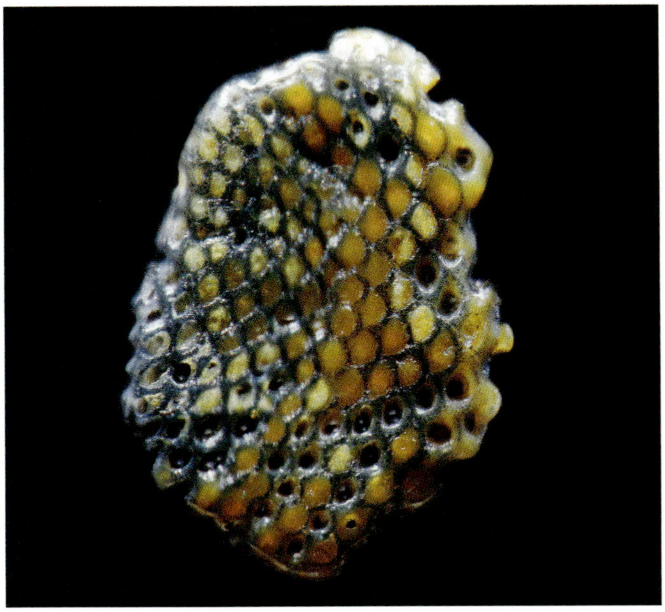

A small bit of weathered coral from Koki Beach, Maui, shows the honeycomb pattern characteristic of coral (magnification 100x).

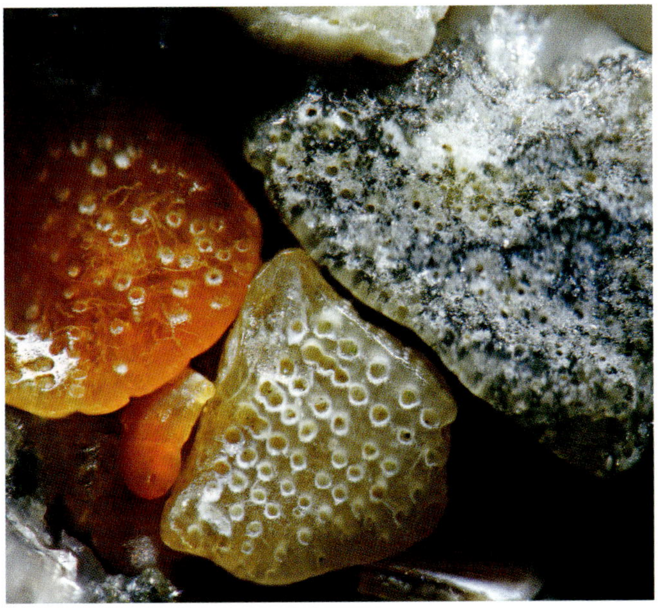

Beautiful bits of colored coral are seen in sand from Lahinch, Ireland (magnification 110x).

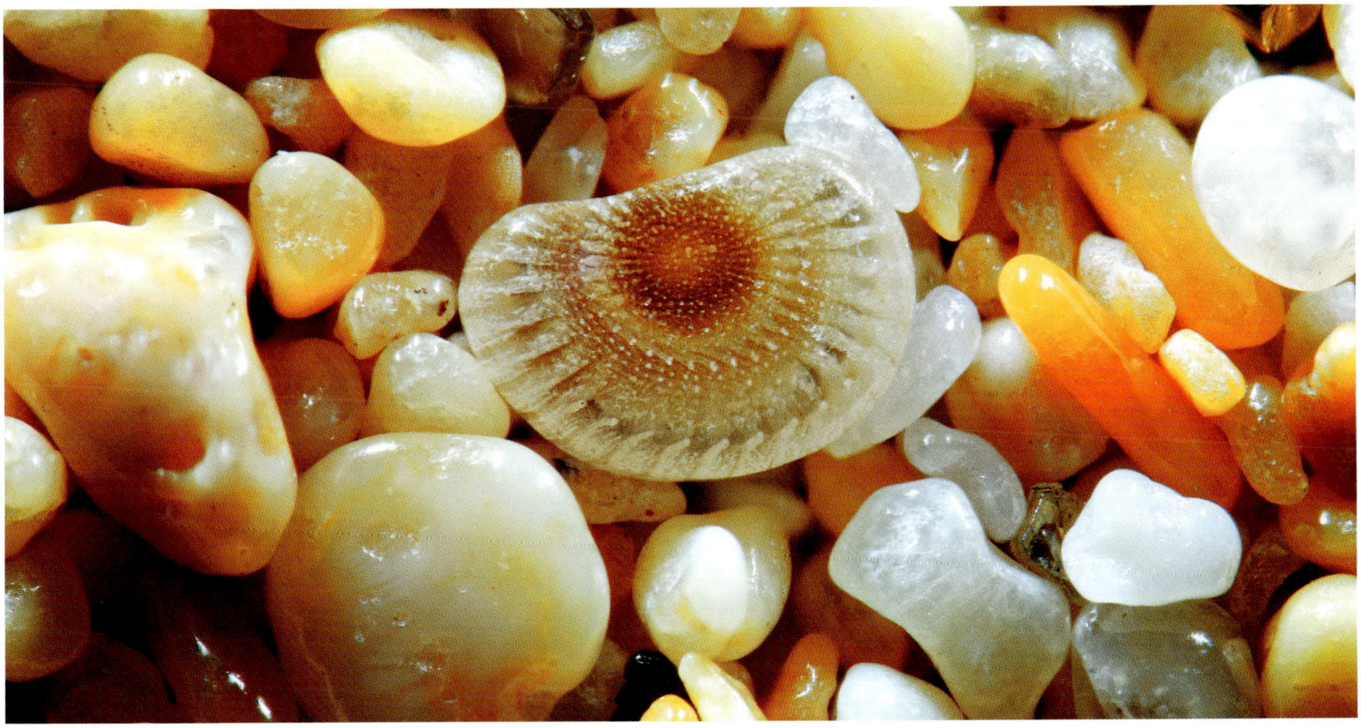

Sand from Makena Beach, Maui, features a mandalalike chip of sea urchin spine among honey-colored sand grains (magnification 90x).

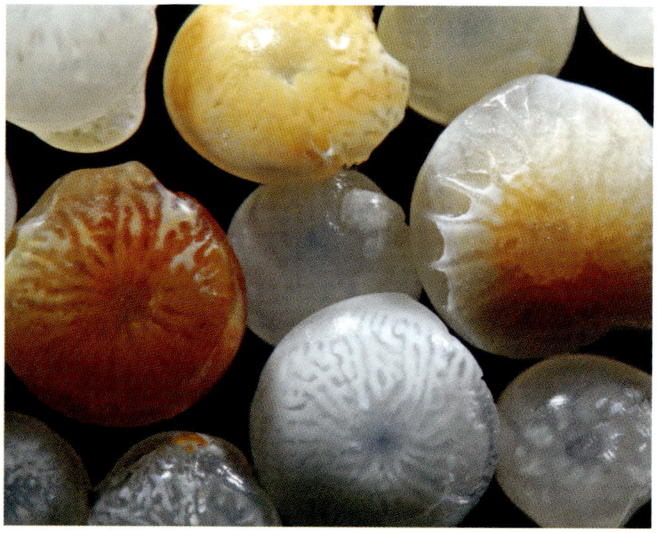

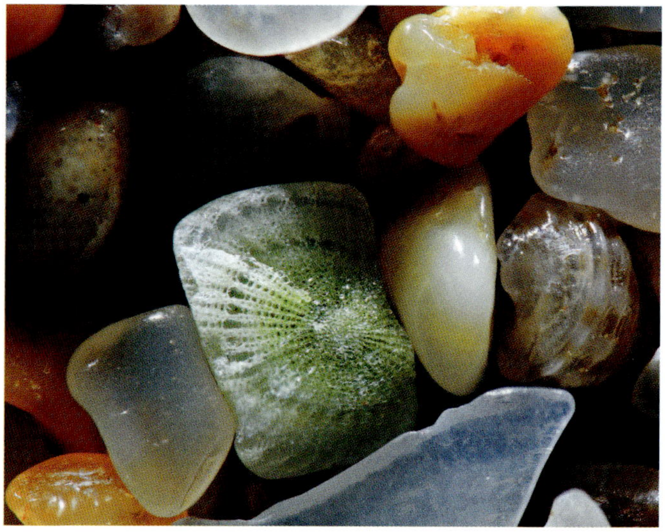

The sand on Wailea Beach, Maui, is filled with biogenic grains. *Left*: A group of beautiful round forams are seen in different stages of erosion (magnification 40x). *Right*: A bit of green sea urchin spine has a fascinating mandala pattern among well-polished sand (magnification 95x).

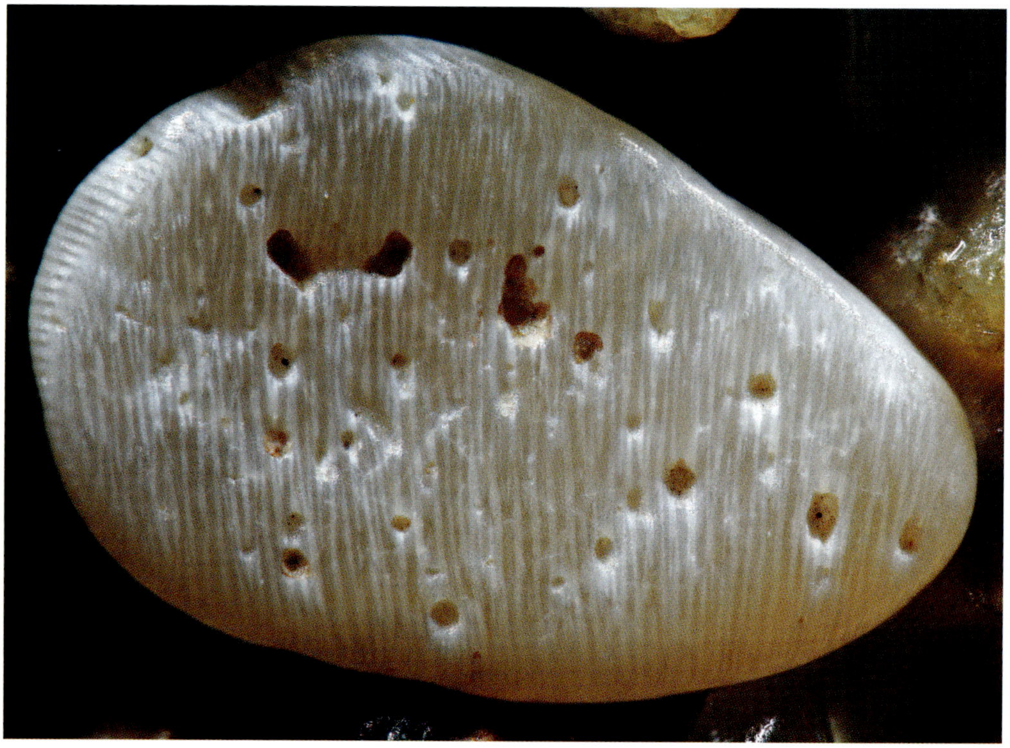

A white shell fragment from Masaya, Nicaragua, shows a striped pattern. The pitting in the shell illustrates the early stages of a shell dissolving, which is determined by the chemistry of the water and the structure of the shell (magnification 125x).

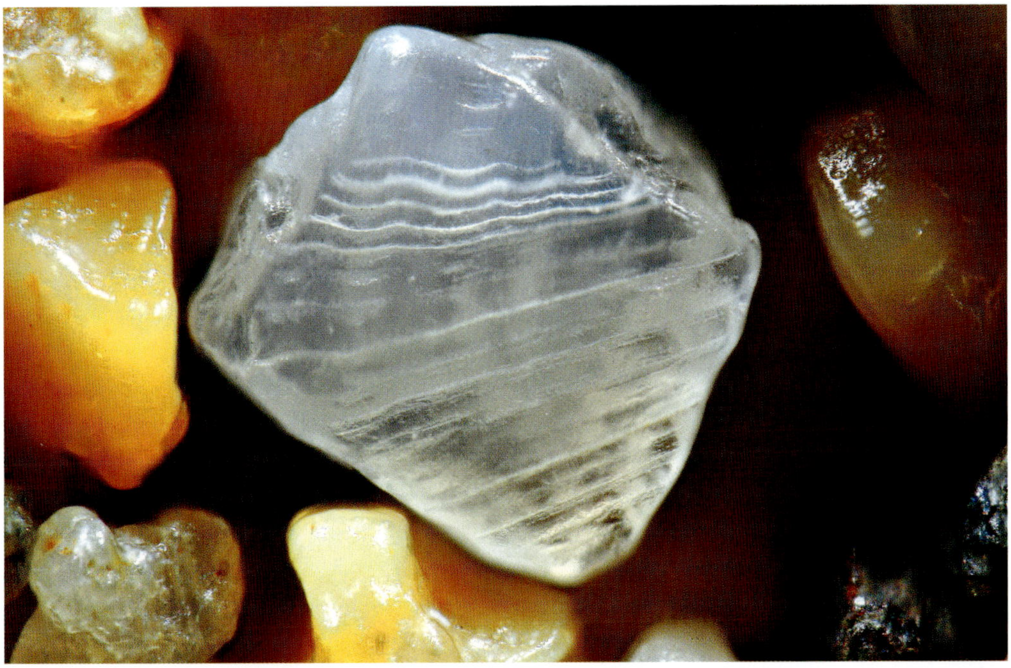

A shell fragment has a wavy striped pattern among rounded and polished sand from Ulua Beach, Maui (magnification 160x).

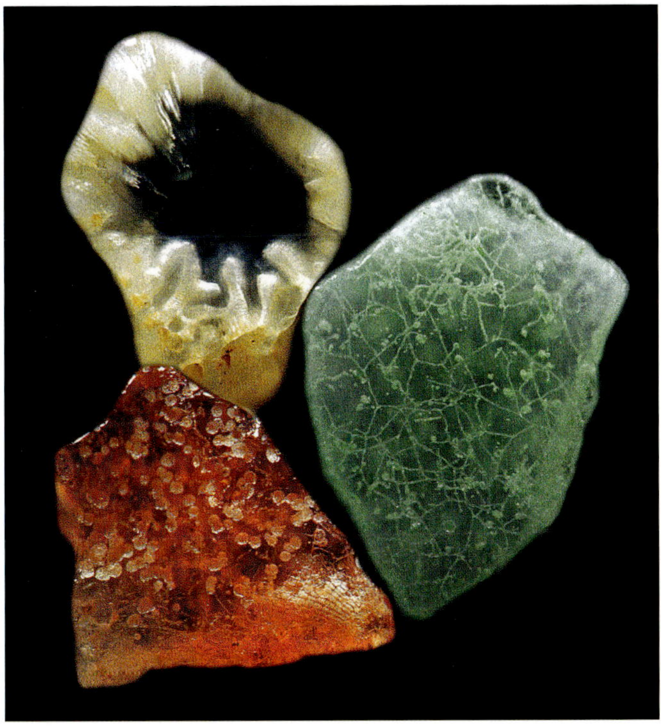

Fine worm tracks mark a green and a pink shell fragment found on a Maui beach. The white sand grain is an extremely eroded foram, showing its internal structure (magnification 90x).

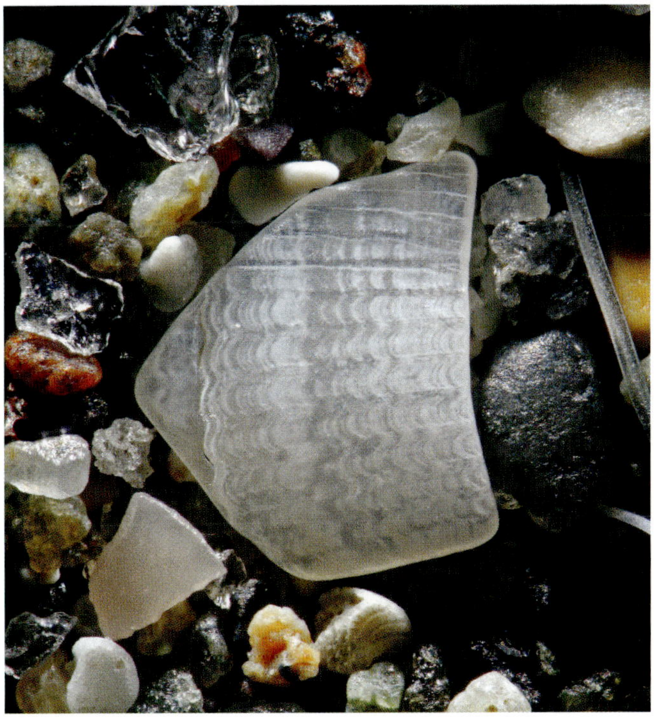

A shell fragment in sand from Zushi Beach, Japan, has a wonderful wavy pattern (magnification 100x).

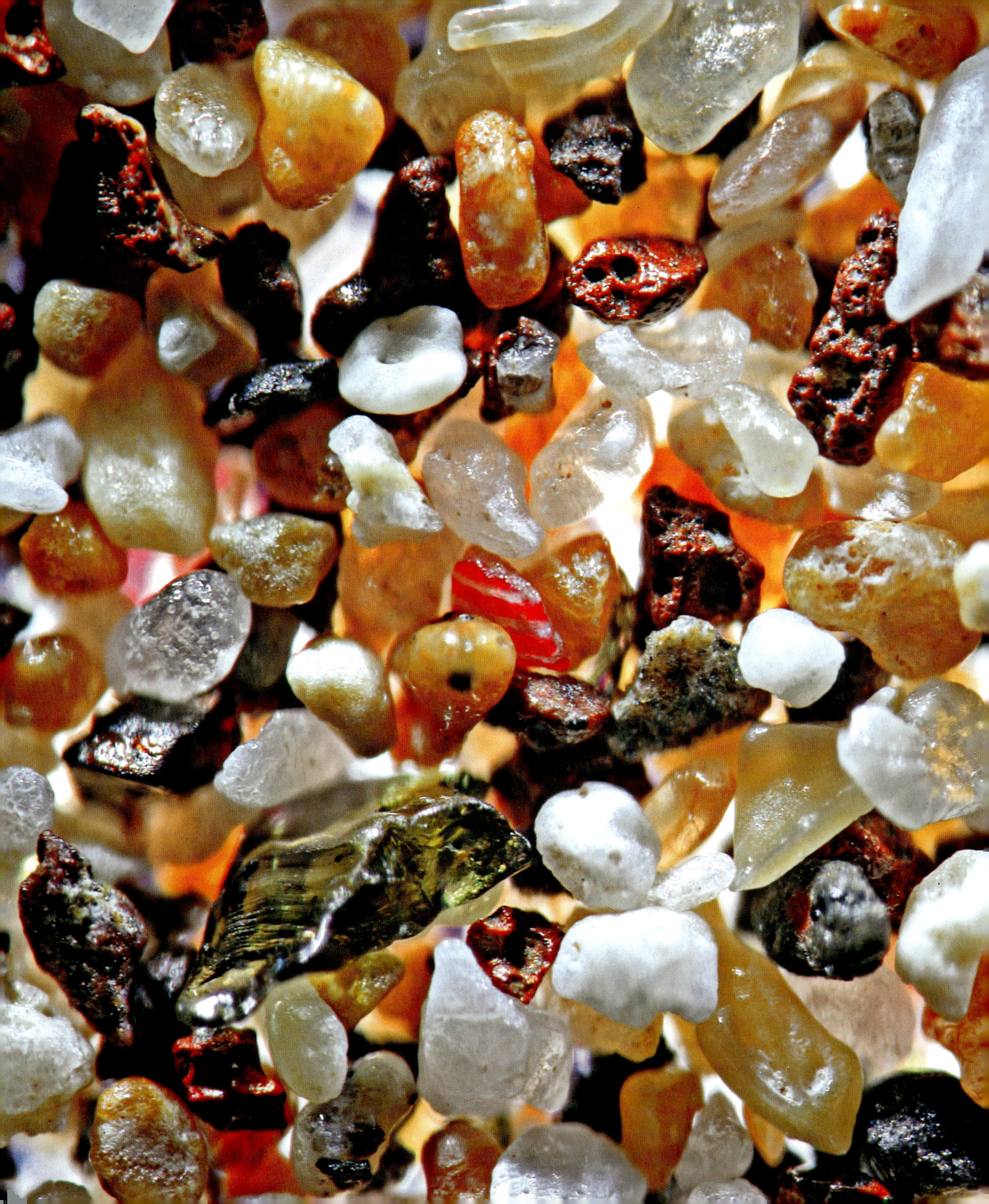

Afterword

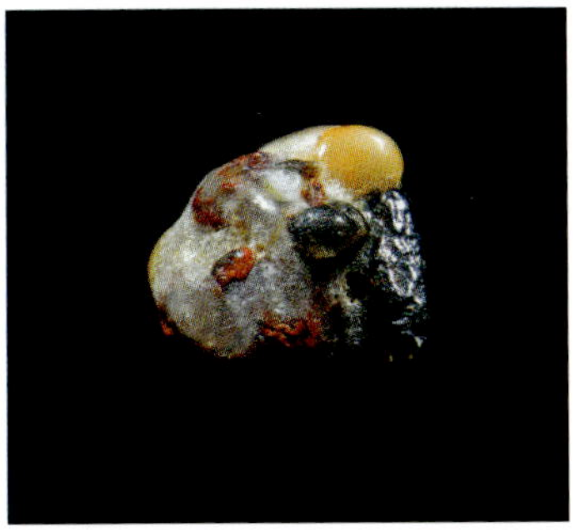
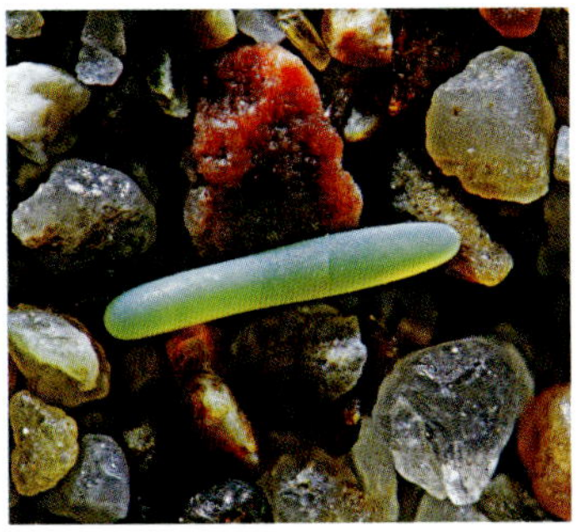
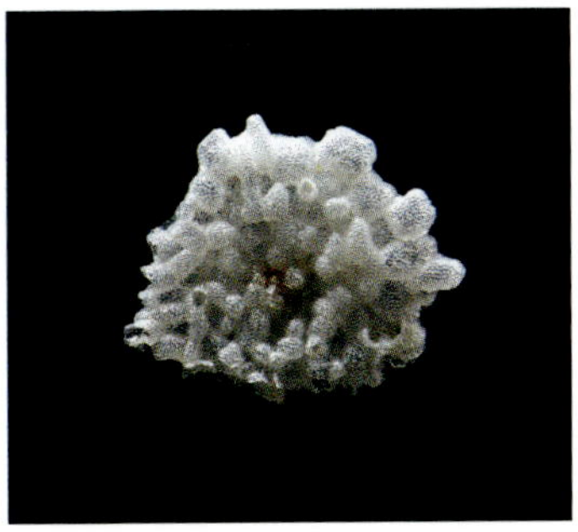

In 1976, when I first began my graduate studies in biology, I had the good fortune to meet Professor Richard Feynman, the great physicist, educator, and Nobel Laureate. David MacDermott, a mutual friend, got the three of us together for dinner one evening at a nice Italian restaurant in Santa Monica. I was a bit nervous because Professor Feynman was the high priest of science to me, and he was arguably one of the greatest scientists of the twentieth century. Among other accomplishments, he cleverly reduced the complex subatomic interactions of quantum physics to a series of simple illustrations called the Feynman Diagrams.

Professor Feynman knew my friend David from Caltech, where David was an artist. David was teaching Feynman how to paint. Feynman's artist friends would teach him art, and he would teach them science in return. He told me at dinner that evening that he didn't think our society was truly in a scientific age because we hadn't yet embraced science in our art and our culture.

Black, white, and pink sand grains from Koki Beach, Maui, Hawaii (magnification 85x).

That statement struck me to the core. It reinforced what I so strongly believe: that science and art are not mutually exclusive, and that they fit beautifully well together. For me, there is little difference between the two. They're both ways of exploring nature and the human condition. They both require creativity, originality, and passion.

I think one of the major differences that does exist between art and science is that art explores nature subjectively, while science attempts to explore nature as objectively as possible. In science, the same question should lead to roughly the same answer, regardless of who asks it. In art, each individual artist will hopefully answer the same question in a unique way. That isn't to say that science leaves no space for subjectivity. Without subjective and intuitive thought, a scientist would never come up with a new hypothesis and science couldn't move forward.

The interplay between art and science forms the basis of the point of view that I have fashioned in my microphotography of sand grains. I express my passion for art, science, and invention in everything I do. As a teacher, I first entice my students by showing them 3-D art through the microscope. I use it as a doorway, as an opening to their innate curiosity; they then become interested and naturally want to learn more.

Almost without exception, people exhibit universal fascination with my images of sand grains through the microscope. One of my greatest rewards is when people come up to me and say, "I never walk along a beach the same way again after seeing your images of sand grains. It's awesome."

I've looked at so many things through the microscope in thirty years that it has shaped the way I see the world. Understanding how things come together on the microscopic level adds to my appreciation of the ordinary, everyday events of life—hopefully it has done the same for you. The universe is endlessly beautiful and praiseworthy.

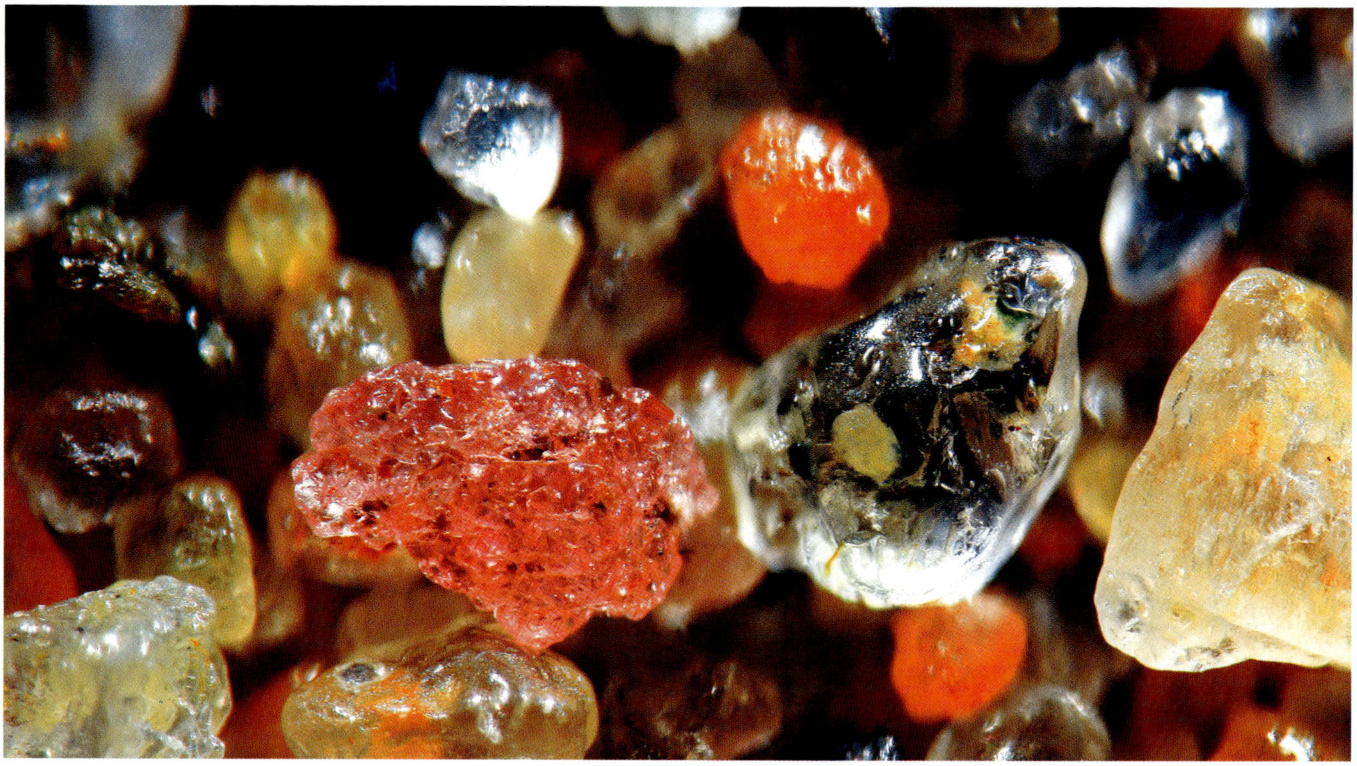

Sand grains from Umhlanga, South Africa (magnification 170x).

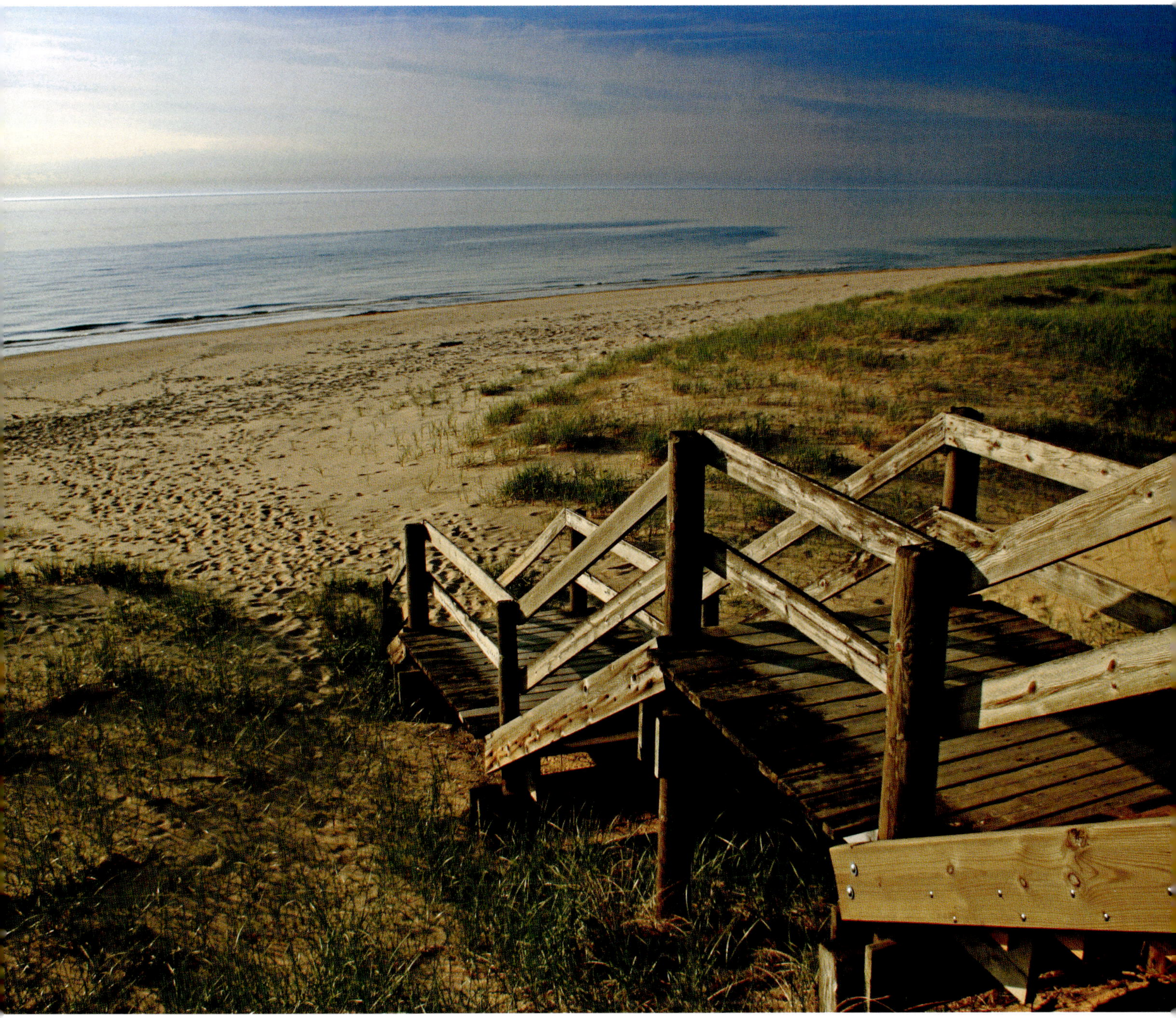

Footprints dot the Lake Michigan shore at sunset. *Noel Powell, Schaumburg, Shutterstock*

Index

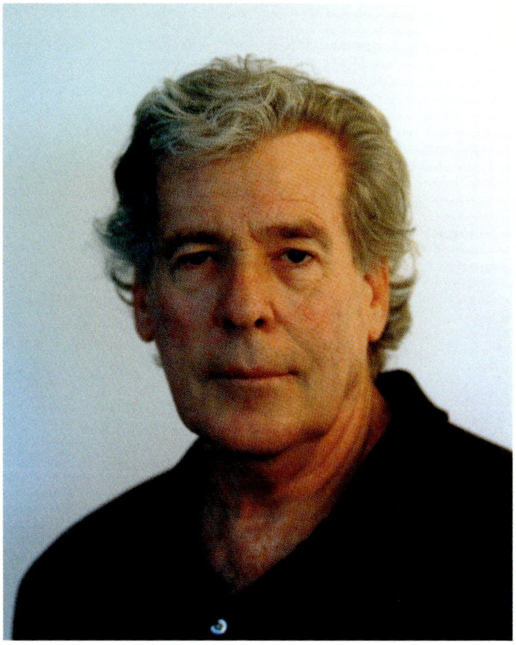

Photo courtesy of author

About the Author/ Photographer

Dr. Gary Greenberg is a visual artist who creatively combines art with science. He has a Ph.D. in biomedical research from University College London and holds seventeen patents for high-definition 3-D light microscopes. He has worked as a photographer, a filmmaker, a biomedical researcher, an academic, an inventor, and an entrepreneur.

Since 2001, Dr. Greenberg has been using his special microscopes to create dramatic landscapes of ubiquitous objects such as grains of sand, flowers, and food. These everyday objects take on a new reality when magnified hundreds of times. Prints of his microscope art are available at www.sandgrains.com.

Dr. Greenberg lives in Haiku, Hawaii.